Lecture Notes in Educational Technology

Lecture Notes in Educational Technology

The series *Lecture Notes in Educational Technology* (LNET), has established itself as a medium for the publication of new developments in the research and practice of educational policy, pedagogy, learning science, learning environment, learning resources etc. in information and knowledge age, – quickly, informally, and at a high level.

More information about this series at http://www.springer.com/series/11777

Dejian Liu · Ronghuai Huang
Marek Wosinski

Smart Learning in Smart Cities

Springer

Dejian Liu
Beijing Normal University
Beijing
China

Marek Wosinski
Arizona State University
Tempe, AZ
USA

Ronghuai Huang
Beijing Normal University
Beijing
China

ISSN 2196-4963 ISSN 2196-4971 (electronic)
Lecture Notes in Educational Technology
ISBN 978-981-10-4342-0 ISBN 978-981-10-4343-7 (eBook)
DOI 10.1007/978-981-10-4343-7

Library of Congress Control Number: 2017936649

Printed on acid-free paper

This Springer imprint is published by Springer Nature
The registered company is Springer Nature Singapore Pte Ltd.
The registered company address is: 152 Beach Road, #21-01/04 Gateway East, Singapore 189721, Singapore

Preface

Smart city is a complex system including smart economy, smart transportation, smart environment, smart learning, smart household life, and smart governance. Creating a smart learning environment in smart city is one of the key elements for the innovative and sustainable development.

Lots of research have been done all over the world to clarify the definitions, dimensions, performances, and initiatives of smart cities. Recently, there is a shift of the focus in these research from technologies oriented on infrastructure to citizen-oriented applications. Education plays an important role in sustainable development and livability of a city, as it promotes creativity and innovation. However, only few studies can be found in smart cities from the educational perspectives. This book aims to discuss smart learning in the background of discussions on smart cities.

Smart learning is an emerging area of interest as a component of other related areas, such as smart technology, smart teaching, smart education, smart e-learning, smart classrooms, smart universities, and smart society. Smart learning comprises not only formal or traditional learning (learning that takes place within a school curriculum), but also informal learning which covers all other forms of learning through informal channels (social media, Internet, MOOCs, game-based learning, and so on);

There are three parts of the book. Part I contains three chapters that introduce the concepts of smart city and smart learning, with the aim to explain the relationship between these two concepts.

Chapter 1, after discussing the definitions and applications of smart cities, is proposing a dual-core framework of smart cities. It is stressing the relationship between creativity of cities and the citizen's livable experience closely related to learning environments in a city. Chapter 1 concludes with the discussion on the relationship between development of a smarter city and smart learning environment.

Chapter 2 aims to analyze smart learning environments in a broader way by integrating both formal and informal learning, as well as physical and virtual spaces in city learning scenarios. After proposing the definition of smart learning environments, author analyzes the six aspects of smart learning environment which are as follows: learning resources, learning tools, learning communities, teaching communities, ways of learning, and ways of teaching. The authors also propose a system model and $TRACE^3$ functional model of smart learning environment. Finally, they analyze the five typical smart learning environments in a smarter city: "self-learning", "inquiry learning", "learning in doing", "learning in working", and "classroom learning".

In Chap. 3, a framework for smart learning is proposed to show the four components of smart learning: learner, supportive technology, core elements of learning scenarios, and logical laws for teaching and learning. Author also describes the four principles and four kinds of supportive technologies. At the end of Chap. 3, the relationship between smart learning and smart city is discussed.

Part II contains four chapters introducing smart learning in various domains.

Chapter 4 investigates the current status of digital campus in China, mainly from the perspective of network construction, digital learning terminals, learning support systems, and information safety. Results reveal that the network access has improved, the learning terminals are enriched, the construction of teaching support service system is unbalanced in different schools, and information safety has been enhanced. Following the analysis of the status, the future trends of digital campus are discussed, and 10 best practices of digital campus in China are introduced.

Chapter 5 is presenting the status of ICT in classrooms in China. Firstly, the trends of ICT in classroom are analyzed. Secondly, the status of multiple types of classroom (multimedia, computer, interactive whiteboard, and tablet computer classrooms) is analyzed. Thirdly, ways to build smart classroom are discussed from classroom equipment, spatial arrangement, teaching application, and cases.

Chapter 6 introduces the development and current status of corporate universities in China, including their background, features, functions, teachers, and the structure of courses. The connection of learning elements in the corporate university and smart learning is also identified with case analysis in this chapter. The online platform of corporate university is introduced by description of its framework, functional module, application, and effective implementation. In the end, the future directions of corporate universities are discussed.

Chapter 7 discusses the developmental stage and main features of smart learning industry in China. The analysis shows that the industry of smart learning develops rapidly in China as a result of increasing market demand, the strengthened support from governmental policies, and the encouragement from capital investors. It cultivates three typical business models: the digitization of traditional regular education, development of learner-centered learning platforms, and creating online schools for professional training in subdivided fields.

Part III of the book contains two chapters discussing the trends of smart learning in smart cities from different perspectives.

Chapter 8 outlines the developmental trends of smart learning in China and offers suggestions for the government how to promote smart learning industry in the long term. The impact of information revolution on smart learning is analyzed from the perspective of learning society, learning behavior, mobile learning, smart city, and learning innovation. The future trends of smart learning are discussed in the context of integration with "Internet +", reconstructing the structural of learning, reconstructing learning environments, reforming online education, and upgrading smart learning.

Chapter 9 reviews some opinions on the role of multicultural perspectives on smart cities proving that just providing technological infrastructure is not enough for success of smart school. In planning of a smart school and organizing smart learning environment, it is necessary to consider local traditions, values, and other cultural factors, as they will determine the motivation of both students and teachers. An important trend in development of smart schools and smart cities will be focus on training of teachers providing knowledge not only on technology, but also on impact of cultural context on learning. This brief analysis may allow the reader to better understand the specificity of Chinese perspective on these concepts, and by connecting the local with the global, we all can more efficiently contribute to the realization of the United Nations Sustainable Development Goals.

Smart learning as element of education plays an important role for the sustainable development of smart cities. This book intends to let reader to understand the relationship of smart learning and smart cities, to know the status of smart learning in China in specific domains of digital campus, digital classrooms, cooperate universities, and smart learning industries. We hope that policy makers, researchers, teachers, practitioners, and graduate students from China will better understand global context of this relationship. We also hope that as this book is also giving Chinese perspective on smart learning and smart cities, it will find interest among the global audiences.

Beijing, China — Dejian Liu
Beijing, China — Ronghuai Huang
Tempe, AZ, USA — Marek Wosinski

Contents

Part III Future Trends: Chinese and Multicultural Perspectives

Part I
Introducing Smart Learning and Smart Cities

Chapter 1
Development of Smart Cities: Educational Perspective

Abstract Lots of research has been done on smart cities to clarify the definitions, dimensions, performances, and initiatives all over the world. Recently there is a shift of the focus in these research from technologies oriented on infrastructure to citizen-oriented applications. Education plays an important role in sustainable development and livability of a city, as it promotes creativity and innovation. However, only few studies can be found on smart cities from the educational perspectives. This chapter, after discussing the definitions and applications of smart cities, is proposing a dual-core framework of smart cities. It is stressing the relationship between creativity of cities and the citizen's livable experience closely related to learning environments in a city. The chapter concludes with the discussion on the relationship between development of a smarter city and smart learning environment.

Keywords Smart city · Smart learning · Smart learning environment

1.1 Introduction

The number of high-density city populations is growing across the world, putting a strain on energy, transportation, water, buildings and public spaces. "Currently 54% of the world's population live in urban areas and the number is forecast to reach 66% by 2050", stated in a report released by United Nations in 2014 (Wolff et al. 2015). Nowadays smart cities are emerging in lots of countries to deal with urban sustainability issues, and to solve the crucial problems in traffic, pollution, city crowding, and poverty by using high technologies (Dameri 2013).

In the initial stage, both academic research and the practice of the smart city focused on how to utilize information technology to improve economic and political efficiency (Röller and Waverman 2001). With the development of smart city, the focus was no longer limited to the hardware facilities, but more attention was paid to the availability and quality of knowledge communication, as well as to social infrastructure (human and social capital). Researchers argued that

D. Liu et al., *Smart Learning in Smart Cities*, Lecture Notes in Educational Technology, DOI 10.1007/978-981-10-4343-7_1

infrastructure construction is the basis of smart city, but a city's **cultural and educational environment** is a driver for city's sustainable development (Caragliu et al. 2011).

With smart city technologies emerging and gradually contributing to a more sustainable and green future, it is becoming apparent that creative skills to deal with these innovations in smart city must be taught to upcoming generations (Wolff et al. 2015). The new citizens will have vital roles in building smart cities to promote all the innovations, who should be hyper connected, creative, entrepreneurs, and also actively participate in the cities' activities and decisions (Ljiljana and Adam 2015).

Education should become a vital part of the future of smart cities (Hall et al. 2000), and learning dimension is becoming more central within smart city discussions. Smart city initiatives should include more investments in training and in continuing education, in order to foster city's learning and innovation capacity (Neirotti et al. 2014).

The most important subjects in the smart city are the citizens; however, they are often ignored (Dameri 2013). Education and learning opportunities for citizens are drivers for a city's sustainability and competitiveness. However, little research has been done on smart city from educational perspective.

Therefore, this chapter intends to explore the educational function in a smart city through reviewing the development of smart city from educational perspective. In Sect. 1.2, the development of smart city definitions will be reviewed to identify the change of emphasis of building a smart city in different cultural and economic contexts. Section 1.3 is introducing the dual-core framework for building smart cities. The relationship between smart learning environment and smart city is discussed in Sect. 1.4.

1.2 Definition of Smart City

1.2.1 Reviewing the Definition

The concept of smart city was born in 1994, but papers on this topic are few until 2010, when "smart" was used by the European Union to qualify sustainability projects and actions in the urban space (Cocchia 2014). Initially, smart city's vision was that government, academia and industry should present information and initiatives in a more technological and informed manner than previously, and the focus was on the significance of new ICT with regard to modern infrastructures (Monfaredzadeh and Berardi 2015).

The typical technology-oriented definition was "an instrumented, interconnected and intelligent city" by Harrison et al. (2010) from IBM. "Instrumentation enables the capture and integration of live real-world data through the use of sensors, personal devices, and other systems; interconnected means the integration of those data into an enterprise computing platform and the communication of such information among various city services; intelligent refers to the inclusion of complex analytics, modeling, and visualization to make better operational decisions" (Harrison et al. 2010).

However, researchers pointed out the disadvantages of technology-oriented strategies, such as risk of social disparities, unequal access to knowledge on ICT usage—the digital divide, spatial polarization and gentrification, citizen lock-in, control and surveillance, high costs, difficulties in integration across different systems, lack of trained staff, etc. (Angelidou 2014).

Another kind of definition of smart city was people-oriented. Paskaleva (2011) viewed smart cities as "people-based, human and progressive in their deployment of digital technologies, not to hardwire themselves, but instead to be socially inclusive in using them to foster good governance and create services capable of improving the quality of life" (Paskaleva 2011).

The advantages of people-oriented strategies include advancement of human capital, advancement of social capital, behavior change, and humane approach. The disadvantages are that not all people have equal access to cyberspace, and the availability of data does not automatically guarantee the enhancement of knowledge and ensure integrity (Angelidou 2014).

Finally, the governance-oriented definition of smart city refers to the administrative and organizational aspects of the city. Smart city is the new concept and new mode to promote the wisdom of urban planning, construction, management and service by using the Internet, cloud computing, big data, geospatial information integration and other new-generation information technology (National Development and Reform Commission 2014). This kind of definition is often promoted by the governments and has more emphasize on the administrative aspects of smart city. It also accentuates the importance of meeting global energy and environmental standards such as energy efficiency and GHG emissions (Mosannenzadeh and Vettorato 2014).

The above three approaches to the definition of smart city mainly reflected the two factors of building a smart city, which are the way cities can steer themselves to achieve this goal of optimization, and the domains that are more critical for a cleverer usage of urban resources (Neirotti et al. 2014). The former refers to the construction of infrastructures of smart cities and the latter refers to the dimensions of smart cities.

1.2.2 Dimensions of Smart Cities

Smart city was developed from digital city, which is a comprehensive, web-based representation of several aspects or functions of a specific real city that includes the dimensions of social, cultural, political, ideological, and also theoretical. Digital city utilized the remote sensing technology, geography information systems and global positioning system to provide support for urban planning, intelligent transport, grid management and services, location based services, urban safety emergency response. With the utilization of technologies, such as Internet of things, Cloud computing, and technologies for digital city, smart city should be able to provide more support for citizen living, urban planning, smart transport, and other dimensions. This section will discuss the dimensions and application of smart cities.

Nam and Pardo (2011) identified and clarified the key dimensions of smart cities, and re-categorized them into three categories of core factors: technology (infrastructures of hardware and software), people (creativity, diversity, and education), and institution (governance and policy). The terms of digital city, intelligent city, ubiquitous city, wired city, hybrid city and information city are more technology oriented (Novotný et al. 2014). The technology factor includes physical infrastructure, smart technologies, mobile technologies, virtual technologies, digital networks, Internet of things, cloud computing, remote sensing technology, geography information systems and global positioning system, etc. Worldwide research on smart city is currently incorporating various aspects of information technology, such as: wireless sensor network node technology, system miniaturization, intelligent wireless technology, communication and heterogeneous network, network planning and deployment, comprehensive perception and information processing, code resolution service, searching, tracking, and information distribution (Albino et al. 2015).

The human factors of smart cities are include in such concepts as creative city, learning city, humane city, and knowledge city, where the focus is on education, learning, human infrastructure and social capital. Education is a critical magnet that makes a city attractive (Nam and Pardo 2011). The institution factor not just includes the supportive policies but also the role of government, the relationship between government agencies and non-government parties, and their governance.

The dimension and application of smart city also include general municipal and business service, intelligent, sustainable buildings and building management (smart building), education, health and social care arena (smart education), energy production and energy efficiency (smart energy, smart lightening), gas, electricity and water smart metering (smart grid), public safety, security and crime prevention, real-time locating services and geographic. Neirotti et al. (2014) provided with a comprehensive understanding of the dimensions of smart city through the elaboration of a taxonomy of pertinent application domains, namely: natural resources and energy, transport and mobility, buildings, living, government, and economy and people.

The most often cited dimensions of smart city appeared in the report of “Smart Cities-Ranking of European Medium-Sized Cities” by Vienna University of Technology, University of Ljubljana, and Delft University of Technology, in which Giffinger (2007) identified the six characteristics as a roof for the further elaboration of smart cities: smart economy, smart people, smart governance, smart mobility, smart environment, smart living.

Smart economy mainly includes factors of urban economy competitive powers such as innovative spirit, entrepreneurship, economic image and trademarks, productivity, flexibility of labor market, international embeddedness, as well as ability to transform. Smart people are not merely described by the level of qualification or education of the citizens but also by the quality of social interactions regarding integration and public life and the openness towards the "outer" world (Balakrishna 2012). The dimension of smart people includes social and human capital factors, the level of qualification, affinity to life long learning, social and ethnic plurality, flexibility, creativity, cosmopolitanism/open mindedness, and participation in public life. As the amount and quality of smart people in a city mainly depends on the education, citizens should be provided with the smart environment to facilitate their learning.

Smart governance assumes participation of citizens in decision-making processes, transparency of governance systems, availability of public services and quality of political strategies. Smart mobility mainly includes the accessibility of local and economic information, availability of information and communication technologies, modernization and sustainability of the transport systems. The dimension of smart mobility includes the transport and ICT factors of local accessibility, (Inter-)national accessibility, availability of ICT-infrastructure, as well as sustainable, innovative and safe transport systems. In Chinese contexts, the word mobility was often understood as infrastructures for mobile communication. In order to avoid misunderstanding, we use smart travelling to represent this dimension.

Smart environment is described by attractive natural conditions (climate, green space etc.), pollution, and sustainable resource management and also by efforts towards environmental protection. Smart living covers most aspects of living qualities such as cultural facilities, health conditions, individual safety, housing quality, educational facilities, touristic attractivity, and social cohesion. Such elements are mainly relevant to people's household lives.

From the analysis of these definitions and applications, we can see that there is a shift of the emphasis for building smart cities from infrastructure or technologies to citizen-oriented application. Infrastructure and technologies construction are the basis for other applications and utilizations of smart city. For citizen-oriented application, the dimensions of creativity, knowledge, sustainability, innovation, and living are all connected with education and learning in smart city. The next section will give a new model for understanding smart cities by considering the core applications and by emphasizing the educational function of a smart city.

1.3 Dual-Core Framework of Smart Cities

The above analysis of the concept and dimensions of smart cities reviewed that the construction of smart cities have evolved from technology-oriented infrastructure construction to people-oriented application construction. The key dimensions and applications of smart city reflected in the aspects of living, mobility, learning, governance, economy, environment, etc., which could also be abstracted as three aspects of technology, people and institution. In the process of constructing smart city, not only the dimensions and applications should be considered, but also the relationship between these dimensions and applications. The following section will discuss the relationship of the core dimensions of smart city, and based on that the dual-core framework of smart city was proposed.

1.3.1 Dual-Core Framework

Xiong (2015), who is the chief scientist for the construction of smart city in China, says that the targets of smart city development are: (1) to build livable, well-being, and informationized cities; (2) to create intensive, efficient economic developing models; (3) to form scientific, reasonable planning and decision-making mechanism.

The first target includes the dimensions of education, environment, traffic, etc. that associated with citizen's livable experience; the second target includes the dimensions of economy, environment, learning, etc. that associated with the development of economy; the third target mainly refers to the dimension of governance. In fact, it is easy to see that the first target is focused on citizen's livable experience and the second and third targets are focused on city's development. If we take citizen's livable experience as the micro level, then the city development and construction is the macro level. Thus, the construction of smart cities on the micro level focuses on creating environments livable for the citizens, while on the macro level focuses on creating innovative developing environment, which are **the two cores of smart cities construction**. The six dimensions and applications of smart city construction can be further analyzed in light of the macro (the city construction) and micro (the citizens) level of smart city construction targets.

On the macro level, the construction of smart cities covers three aspects: The first one is smart economy, which involves urban enterprise creativity, the "internet +" economic form and its employment entrepreneurship atmosphere. The second aspect refers to smart environment which covers the planning of urban buildings, green energy and green city. The third aspect relates to smart governance (management) which mainly involves urban service policies, openness and transparency of data, as well as the popularization of e-government. In another words, the smart features of a city could be represented by "the vitality for urban innovation" which includes the three characteristics of smart economy, smart environment and smart governance.

On the micro level, the construction of smart cities could also be considered in three aspects. The first one is smart mobility (travelling) which aims at providing the citizens with convenient transport facilities and efficient, ubiquitous access to internet. The second aspect includes smart living, which refers to people's sense of security of urban life, medical and health conditions, and the citizen's sense of happiness. The third aspect refers to smart people (learning), that should provide skills that the citizens should master in the 21st century. This can be achieved by accepting inclusive and technology integrated education.

Therefore, smart city construction mainly consists of city-oriented macro level construction and citizen-oriented micro-level construction. The smart features that citizens can feel is "citizens' livable experiences" which cover smart travelling, smart living and smart learning. The concept of "smart citizens" in the context of "smart city" ultimately refers to smart learning.

It could be indicated from the above analysis that the mean of smart cities construction is diversified with many aspects centering on the two cores: "Citizen's Livable Experiences" from the micro-level and "City Innovation Capacity" from macro-level. Therefor, we proposed the dual-core framework of smart city, as shown in Fig. 1.1. On the left of Fig. 1.1, the core of Citizen's Living Experience includes smart travelling, smart learning, and smart living which together provide bountiful living experience for citizen. Smart learning takes 21st century skills, inclusive education, infusing ICT into education, etc. into consideration; smart living takes urban security, medical and health care, civil happiness, etc. into consideration; smart travelling takes convenient traffic, efficient access, ubiquitous network access, etc. into consideration. On the right of Fig. 1.1, City Innovation Capacity includes smart environment, smart economy and smart governance which together provide the vitality of innovation. Smart environment takes green building, green energy, green urban plan, etc. into consideration; smart governance takes service policy, transparency and open data, widespread use of digital government etc. into consideration; smart economy takes entrepreneurial creativity, internet + economic pattern, employment and venture opportunities, etc. into consideration.

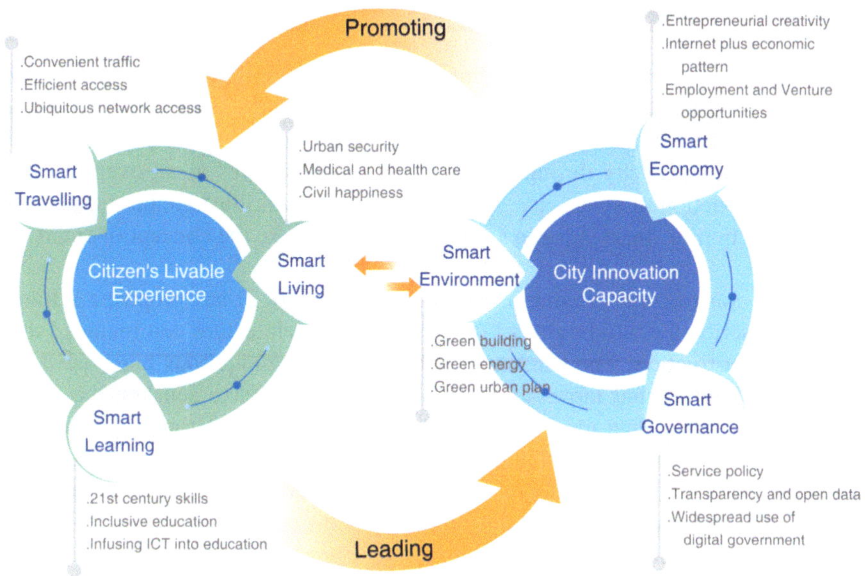

Fig. 1.1 Dual-core framework of smart city

The two cores of smart cities construction serve as the basic drive to promote a city's development, the main aim of which is to improve a city's benign operation and to solve the problems facing "smart city construction" dilemma. Advocating "smart learning" plays a cultural leading role for stimulating vitality for urban innovation and it also provides scientific support for citizens' livable experiences. In fact, it is the ultimate target of smart city construction to let humans have better urban living environments. The two cores that represent features of a smart city indicate that smart learning is the fundamental driving force to enhance citizens' wisdom and the basic solution to improve people's livable experiences.

Smart city definitions also highlight important aspects of sustainability, such as the need for responsible resource management and energy efficiency (Monfaredzadeh and Berardi 2015). In the course of construction, it is very essential to develop a scheme that integrates every aspect, combing city planning with energy planning as it is necessary to stress innovation and sustainability. Innovation is the driving power of urbanization and is the basis for sustainability, such a principle does not only apply to urbanization in a single area, but all over the world. According to the two cores of smart city, smart learning is the drive for city innovation vitality, which also plays a key role for the sustainability of a city. A city becomes "smart" when it is adaptive, learning, autonomous, self-repairing, and robust (Novotný et al. 2014). Therefor, smart learning is the foundation of smart city construction, and the relationship of smart learning and smart city will be future discussed in Chap. 3.

1.3.2 Construction of Smart Cities

The dual-core framework of smart city indicates that the two cores of civil livable experience and urban innovative capability should be considered in the process of smart city construction. The understanding of the relationship of the six different dimensions of smart city through dual-core framework gives us an overall picture for the construction of smart city. The following part will illustrate the methods for the construction of smart cities in China, which will make the concept of smart city concrete.

For the construction of smart cities, Balakrishna (2012) has proposed a framework to describe the building blocks of smart cities architecture that included three levels of construction. The fundamental level was the ubiquitous high-speed network infrastructure and large-scale instrumentation-pervasive sensors; the second level dealt with the data management, ambient intelligence, autonomous decision; the third level was the smart applications and services, such as smart environment, smart economy, smart people, smart governance, smart living, and smart mobility.

In China there are also three routes for the construction of smart cities. The first route centers on "data" construction, in which internet companies (such as Baidu, Alibaba, and Tencent) sign contracts with local governments to build smart cities. The main aim and core value of such operations are to realize comprehensive analysis and application of data. For example, on 15th May 2015, the Shanghai Government signed a strategic co-operation frame agreement with Alibaba Group, who is in charge of providing technological supports to cloud computing and big data industries in Shanghai as well as providing resource support of brands, technologies, service and training. The aim of this route is to provide personalized service by analyzing big data. The problem is that such applications cannot offer effective supports for the industrial economy.

The second route focuses on "integration" construction, which integrates applications from various sectors. For example, Data China's business covers the entire industrial chain: design, business tackling, development and implementation as well as online operation and maintenance. The enterprise defines itself to be a full smart city service operator which offers services for all sectors of smart city applications. The aim of this route is to achieve better utilization of existing information investments, and key point lies in integration of multi-systems which may result in the impossibility of fundamentally eliminating "information island".

The third route centers on "industrial" construction, which refers to the construction and service in specific and vertical industries, especially for traditional middle-size and small-sized enterprises that make profits by way of providing deep solutions and specialized applications in professional fields. Such construction aims at specializing industrialized development in a specific sector. However, such a scheme may block the city's general planning of operation and construction.

With the deepening specification of "internet +" and "Industry 4.0" applications, people express their multi-layer, diversified needs for comprehensive smart city constructions and also hope that the construction and development of smart cities are compatible with their daily needs. Based on studies of existing constructing schemes, it can be concluded that smart city construction usually covers three aspects: urban infrastructure construction, local industries construction and economic development construction. Such constructions, by far, may not completely satisfy the people's living needs, which diverges from one of the key objectives of smart city—"creating livable and happy living environments" to a certain level, which brings deeper-level constructing problems.

1.4 Smart Cities and Smart Learning Environments

Since the initiatives of constructing smart city, the urban infrastructure has been improved continuously all over the world, but city's ability to raise innovation was a problem that lots of cities are currently facing. Building smart learning environment in smart city to promote both formal and informal learning for educating future ready citizen, is one of the methods to solve this problem. In order to meet people's formal and informal learning needs, smart learning environment was proposed to build the learning place or virtual space that can aware learning scenarios, identify the characteristics of learners, provide appropriate learning resources and convenient interaction tools, automatically record the learning process and evaluate learning outcomes in order to promote the effective learning of the learners (Huang et al. 2012). If citizens are to use and even innovate their own sustainable smart city solutions using big urban data, then they must have a good understanding of sustainability issues and possess the essential skills for using large complex datasets (Wolff et al. 2015). With the fast-growing smart cities, it is urgent to improve the learning environment, and expand quality learning resources to match the learning needs of digital citizen. Smart learning could happen in school education, home education, community education, social education, and all these places and virtual space could be regarded as learning environments in a city.

China has paid much attention to encourage all people for innovation and venture, especially the new graduates. Innovation and venture are not only the focus of central government and the key point for urban upgrading, but also put a strong need for optimizing learning environments in a city. Nowadays, the society has already stepped into a new era that deeply blended with spirit world, physical world, and information world (Huang et al. 2013). In such circumstance, many challenges in the planning, implementation, management, operations, and service existed in the process of constructing smart city. The building of smart learning environments and advocating of smart learning could provide a persistently renewing depository of wisdom, technical and talents to meet the needs of social development, to address the challenges of urban development.

A smart city is also a learning city, which improves the competitiveness of urban contexts in the global knowledge economy (Plumb et al. 2007). The competitiveness of a city depends not only on the infrastructure, but more on construction of soft power, such as intellectual capital and social capital. The construction of smart city should be carried on according to the dual-core of "civil living experience" and "urban innovative capacity" of smart city. Smart learning is an important support to bridge educational systems and citizen living experience. The building of smart learning environments for citizen will provide individuals more opportunities to learning easy, engaged and effective, and therefore provide wisdom into the creativity of the whole city.

The improvement of city learning environment is a long-term, and continued process. The nature of smart learning includes: (1) the awareness of learning environment to provide learners personalized and context aware learning, (2) the adaptivity of learning content to provide what learners want to learn and need to learn according to their personality, (3) the equity for all to send quality learning resources to all who want to learn by utilizing emerging technologies, (4) the organic integration of education system elements of learners, teachers, contents, and medias. Smart learning environment will be built to support new model of teaching and learning to inspire the wisdom of citizen, to bring more fair and opportunities to all people lived in smart cities.

Update of urban governance, promotion of public literacy, and ability of attracting young people, etc. has become an important aspect of a city's soft power. A key indicator for deciding the success of a city is the ability of attracting young people, and the ability to keep local young people, as the youth is the main innovation driver of a city. Therefore, young people's learning needs and development needs should be addressed, and optimizing smart learning environment in a city is one of the ways to meet their needs.

In order to utilize smart learning in upgrading of the city's soft power, the government should integrate smart learning environment into the design of the smart city, and to use smart learning as one of the indicators for assessing smart city. Therefore, smart learning environment and smart learning in the context of smart city should be further explored, what is discussed in Chaps. 2 and 3 of our book.

References

Albino, V., Berardi, U., & Dangelico, R. M. (2015). Smart cities: Definitions, dimensions, performance, and initiatives. *Journal of Urban Technology, 22*(1), 3–21. doi:10.1080/10630732.2014.942092

Angelidou, M. (2014). Smart city policies: A spatial approach. *Cities, 41*, S3–S11. doi:10.1016/j.cities.2014.06.007

Balakrishna, C. (2012). Enabling technologies for smart city services and applications. In *Proceedings—6th International Conference on Next Generation Mobile Applications, Services, and Technologies, NGMAST 2012* (pp. 223–227). doi:10.1109/NGMAST.2012.51

Caragliu, A., Del Bo, C., & Nijkamp, P. (2011). Smart cities in Europe. *Journal of Urban Technology, 18*(2), 65–82. doi:10.1080/10630732.2011.601117

Cocchia, A. (2014). Smart and digital city : A systematic literature review. In R. P. Dameri & C. Rosenthal-Sabroux (Eds.), *Smart city, progress in IS* (pp. 13–43). Springer International Publishing Switzerland. doi:10.1007/978-3-319-06160-3

Dameri, R. P. (2013). Searching for smart city definition: A comprehensive proposal. *International Journal of Computers & Technology, 11*(5), 2545–2551.

Giffinger, R., Fertner, C., Kramar, H., Kalasek, R., Pichler-Milanovic, N., & Meijers, E. (2007). *Smart cities: Ranking of European medium-sized cities*. Final Report, Centre of Regional Science, Vienna UT.

Hall, R. E., Braverman, J., Taylor, J., & Todosow, H. (2000). The vision of a smart city. In *2nd International Life Extension Technology Workshop* (p. 28). Paris.

Harrison, C., Eckman, B., Hamilton, R., Hartswick, P., Kalagnanam, J., Paraszczak, J., & Williams, P. (2010). Foundations for smarter cities. *IBM Journal of Research and Development, 54*(4). doi:10.1147/JRD.2010.2048257

Huang, R., Yang, J., & Hu, Y. (2012). From digitalized environment to smart environment: The reform and trends of learning environment. *Open Education Research, 18*(1), 75–84.

Huang, R., Chen, G., Yang, J., & Loewen, J. (2013). The new shape of learning: Adapting to social changes in the information society. In R. Huang & J. M. Spector (Eds.), *Reshaping Learning SE–1* (pp. 3–42). Springer: Berlin, Heidelberg. doi:10.1007/978-3-642-32301-0_1

Ljiljana M., Adam S. (2015) Building a Gamified system for caputring MOOC related data: Smart city learning community as its most precious source of intangible cultural heritage. *2015 International Conference on Culture and Computing* (*Culture Computing*, pp. 175–182).

Monfaredzadeh, T., & Berardi, U. (2015). Beneath the smart city: Dichotomy between sustainability and competitiveness. *International Journal of Sustainable Building Technology and Urban Development, 6*(3), 140–156. doi:10.1080/2093761X.2015.1057875

Mosannenzadeh, F., & Vettorato, D. (2014). Defining smart city: A conceptual framework based on keyword analysis. *Journal of Land Use, Mobility and Environment, 6*, 683–694.

Nam, T., & Pardo, T. A. (2011). Conceptualizing Smart City with Dimensions of Technology, People, and Institutions. In *Proceedings of the 12th Annual International Digital Government*

Research Conference: Digital Government Innovation in Challenging Times (pp. 282–291). doi:10.1145/2037556.2037602

National Development and Reform Commission. (2014). *Guidance on promoting smart city development (Publication No.[2014]1770).* Beijing: China.

Neirotti, P., Marco, A. De, Cagliano, A. C., Mangano, G., & Scorrano, F. (2014). Current trends in smart city initiatives: Some stylised facts. *Cities, 38*, 25–36. doi:10.1016/j.cities.2013.12.010

Novotný, R., Kuchta, R., & Kadlec, J. (2014). Smart city concept, applications and services. *Telecommunications System & Management 3*(2). doi:10.4172/2167-0919.1000117

Paskaleva, K. A. (2011). The smart city: A nexus for open innovation? *Intelligent Buildings International, 3*(3), 153–171.

Plumb, D., Leverman, A., & McGray, R. (2007). The learning city in a 'planet of slums'. *Studies in Continuing Education, 29*(1), 37–50.

Röller, L. H., & Waverman, L. (2001). Telecommunications infrastructure and economic development: A simultaneous approach. *American Economic Review*, 909–923.

Wolff, A., Kortuem, G., & Cavero, J. (2015). Towards smart city education. In *2015 Sustainable Internet and ICT for Sustainability, SustainIT 2015*. doi:10.1109/SustainIT.2015.7101381

Xiong, Z. (2015). Promoting the construction of smart cities by utilizing ICT service industry. *Nation Governance., 18*, 004.

Chapter 2
Contexts of Smart Learning Environments

Abstract Considering the importance of education and learning for the development of creativity and sustainability of a city, smart learning environments (SLE) could be regarded as a response to the needs of the new knowledge society. This chapter aims to analyze smart learning environments in a broader way by integrating both formal and informal learning, as well as physical and virtual spaces in city learning scenarios. After proposing the definition of smart learning environments, we analyze the six aspects of smart learning environment which are: learning resources, learning tools, learning communities, teaching communities, ways of learning and ways of teaching. We also propose a system model and TRACE^3 functional model of smart learning environments. Finally, we analyze the five typical smart learning environments in a smarter city: "self-learning", "inquiry learning", "learning in doing", "learning in working", and "classroom learning".

Keywords Smart learning environments · Smarter city · Digital learning environments · Learning scenario

2.1 Introduction

We live in an information society where the creation, distribution, use, integration and manipulation of information has a significant impact on economic, political, and cultural development of the city. With the expansion of the digital world every sector of the society has been influenced by information technology. The digital world is a combination of the real world, and virtual world, thus, people's life style and ways of working have changed as well as the ways of learning. Nowadays, people learn everywhere, not only formal learning in classroom, but also informal learning at home and society. The spaces of formal and informal may be physical places or virtual environments, which could be called learning environments in cities. Learning environments in smart cities include school learning spaces, home learning spaces, community learning spaces, working places, learning stadiums, and other virtual learning spaces. The aim of smart learning environments is to

D. Liu et al., *Smart Learning in Smart Cities*, Lecture Notes
in Educational Technology, DOI 10.1007/978-981-10-4343-7_2

increase the public opportunities to learn, to improve the scientific literacy and knowledge acquisition capacity of citizens, and to upgrade the city's soft power.

With the development of information technology, the learning ways of mobile learning and digital learning became increasingly universal, digital learning resources constantly became richer, and the demands for personalized and smart learning became stronger. Smart learning environments should be built to provide smart learning opportunities for different kinds of people to meet the diverse learning needs. In a smart learning environment, learners can learn at any time, any place, any way, and at any pace. This kind of learning environment can also support learners with the experience of easy, engaged and effective learning. In other words, smart learning environments engage and integrate formal and informal learning in order to create autonomous adaptive learning environments for supporting individual learners with real-time and seamless learning experiences in ubiquitous settings (Kinshuk et al. 2016).

The term of future learning environment (Sevindik 2010), smart classroom (Lui and Slotta 2014), learning spaces (Long and Ehrmann 2005), smart campus (Kwok 2015), technology-enhanced learning (Chan et al. 2006), etc. are similar to smart learning environments, as all these concepts to refer to utilizing information technology to build for learners a personalized and adaptive learning environment. Hwang (2014) mentioned that technology-enhanced learning can be perceived as smart learning systems. In fact, smart learning environments are integration and combination of lots of different systems, such as intelligent tutoring, learning analytics, educational data mining, adaptive learning, and personalized learning (Yang et al. 2015).

However, the literature on smart learning environments indicates that the research of smart learning environments mainly focused on the technologies and pedagogies, and only few research had been done from a citizen's learning perspective. Therefore, the aim of this chapter is to analyze the definition, elements, functions, and scenarios of smart learning environments in a broader way by integrating both formal and informal learning in physical and virtual spaces in city learning scenarios.

The following sections will identify the definition of smart learning environments, as well as the differences between smart learning environments and digital learning environments. The learning modes in the smart learning environments of school, home and society will be described in Sect. 2.3. The constituent elements and functional model will be proposed in Sect. 2.4. The typical smart learning environments of self-regulated learning, inquiry learning, learning in doing, learning in working, and classroom learning will be discussed in Sect. 2.5.

2.2 Definition of Smart Learning Environments

With the development of ICT in education and considering scaling up the innovation of technology enhanced learning, researchers begun to conceptualize how learning environments can be made more effective, efficient, and engaging on a large and sustainable scale (Spector 2014). The concepts of next generation learning space, smart learning environment, and smart classroom emerged. Learning space is a new emerging research area, with the aim to promote independent, flexible, and engaged learning by providing learner with appropriate technology and pedagogy (Huang et al. 2015). Smart Learning Environments (SLE) are defined as physical environments that are enriched with digital, context-aware and adaptive devices, to promote better and faster learning (Koper 2014). With technology support, smart classrooms become places where teachers and students could have rich and immersive teaching and learning experiences that they have never experienced before (Li et al. 2015). Hwang (2014) presented the definition and criteria of SLE from the perspective of context-aware ubiquitous learning. He also introduced a framework to address the design and development of SLE to support both online and real-world learning activities.

Zhong and Zhang (2006) presented that smart learning environments were a learner-centered intelligent, open, integrated digital virtual reality learning space. He maintained that the core elements of smart learning were corresponding equipment, tools, technology, media, textbooks, teachers, and students, etc., and all these core elements should be organized according to constructivist learning theory, blended learning theory and modern teaching theory. Chin (1997) argued that "smart learning environments were a learner-centered environment based on the application of information and communication technology, with the following characteristics: ① Adapt to different learning styles and learning abilities of learners; ② Provide support of lifelong learning of learners; ③ Provide support of the development of learners." Based on the above points of view and the analysis of the developmental trends of technology-enhanced learning, smart learning environments should have the following characteristics:

1. Smart learning environments should integrate physical environment and virtual environment. In the smart learning environment, the perceptive, monitoring and regulating functions of physical environment are further enhanced. The application of the technology of augmented reality create the seamless integration of virtual environment and physical environment.
2. Smart learning environments should provide better learning support and services according to the individual characteristics of learners. Smart learning environments emphasize the process record, personalized assessment, and evaluation of effects and content delivery of learners' learning. According to the learner model, it plays significant role in planning, monitoring and evaluation in the development learner's learning capabilities.
3. Smart learning environments support in-campus learning and off-campus learning, formal learning and informal learning. The learners in this situation are not only the campus learners, but also all people that have requirements of learning in their work.

Therefore, we regard smart learning environments as the learning place or an activity space that can aware learning scenarios, identify the characteristics of learners, provide appropriate learning resources and convenient interaction tools, automatically record the learning process and evaluate learning outcomes in order to promote the effective learning of the learners.

Smart learning environments are the high-end form of common digitalized learning environments, which are the inevitable result of the development of educational technology. There are significant differences in six aspects of learning resources, learning tools, learning communities, teaching community, learning methods and teaching methods between smart learning environments and common digitalized learning environments, as shown in Table 2.1.

Table 2.1 The comparison of digitalized learning environments and smart learning environments

	Digital learning environment	Smart learning environment
Learning resources	(1) Digital resources are based on rich media; (2) Online access becomes the mainstream; (3) Users select resources	(1) Digital resources are independent of the devices; (2) Seamless connection or automatic synchronization becomes fashionable; (3) on-demand resources
Learning tools	(1) All-function in one tool, systematized tools; (2) Learners judge the technology environment; (3) Learners judge the learning scenarios	(1) Specialized Tools are specialized and miniaturized tools; (2) Technology environment is automatically sensed; (3) Learning scenarios are automatically recognized
Learning community	(1) Virtual community is focused on online communication; (2) Self-selected community; (3) Restricted to information skills	(1) Combined with the mobile interconnected real community to communicate anytime and anywhere; (2) Automatically matched communities; (3) Depends on media literacy
Teaching community	(1) Difficult to form a community, which is highly dependent on experience; (2) Make the regional community possible	(1) Automatical ~ ly form community, which highly concerns about the users' experience; (2) Make the cross-regional community fashionable
Learning methods	(1) Focus on individual knowledge construction; (2) Focus on low-level cognitive objectives; (3) Unify evaluation requirements; (4) Interest becomes the key to the diversity of learning methods	(1) Highlight the knowledge construction of community collaboration; (2) Focus on high-level cognitive objectives; (3) Multiple evaluation requirements; (4) Thinking becomes the key to the diversity of learning methods
Teaching methods	(1) Emphasize resource design and explanation; (2) Emphasize summative evaluation of the learning outcomes based on the learners' behaviors; (3) Emphasize observation of learning behaviors	(1) Emphasize activity design and guidance; (2) Adapt the evaluation of learning outcomes based on the cognitive characteristics of learners; (3) Intervene in learning activities

2.3 Learning Modes in Different Smart Learning Environments

In the ideal smart learning environments, every learner holds an intelligent mobile device (such as iPad) in his/her hand. Its screen is of the size close to printed textbooks. It can imitate all functions of printed textbooks, such as taking notes, inserting bookmarks, making markings and annotations, etc., and has the effect of turning papers as with printed textbooks. This type of "textbook" equipped with intelligent mobile devices is called e-textbook. The contents of e-textbooks are presented via multimedia. The links between knowledge points based on semantic relations can realize the personalized presentation of knowledge contents. Binding with the learning progress of the learners, e-textbooks can realize the synchronized cloud service of learning data, which can record the learning process of the learners, intelligently analyze the learning outcomes of the learners, present the analysis results in graphs and provide guidance and help to the learning of the learners in accordance with the viewpoints of teachers. For campus students, smart learning environments will make their learning in school, at home and in the community full of intelligence.

In the learning on campus, teachers can make use of augmented reality technology to present a variety of authentic learning scenarios, so that students can experience immersive learning objects to enhance their learning interests and motivations. According to the learning records of students in the system, teachers emphasize on the instruction of the knowledge points that students have difficulty and use the rich learning resources to design a variety of learning activities. They can flexibly control the learning terminals and provide real-time delivery of relevant learning resources with an integrated classroom control system. They can make fast grouping according to the characteristics of learners, so as to facilitate the organization of collaborative learning in class.

Students can use the convenient interactive tools provided by the system to interact with peers and teachers. They can also use the built-in voting machine to have real-time interaction with teachers so that teachers can obtain their feedbacks at the first time to adjust instruction based on the feedback information in time. Smart learning environments can provide intelligent support of instructional design to help the teachers with the instructional design in the classroom. It can automatically correct and analyze the homework and test papers submitted by students. It also provides social networking tools for facilitating the contact between teachers and students, and among students. This includes synchronous communication tools such as QQ, MSN, etc. and asynchronous communication tools such as micro-blog, virtual learning communities, etc.

In learning at home, students can use e-textbooks to preview the lessons and complete the assignments at home. While previewing the lessons, they mark out the important points and try to complete the assignments after the preview. The system can automatically give the feedback of the results of the assignments, offer tips and answers to difficult problems and provide the structure chart of the relationship of knowledge points in accordance with the main and secondary knowledge points. The system can record the situation of the completion of students' homework, so teachers can provide targeted instruction and personalized guidance to the students according to the statistics. Smart learning system provides one-on-one counseling function to facilitate the students to "call" the teachers when they need learning guidance. Parents can know about the leaning progress on campus of their children from the learning records provided by e-textbooks, and their electronic signatures can be conveniently transmitted the school management system via e-textbooks.

In the social learning, smart learning environments can perceive the location of the learner. SLE will deliver the learning resources related to the environment of the learner according to their location and the learning style, so as to achieve adaptive ubiquitous learning. In some circumstances, it can divide the learners into groups according to their locations. For instance, it can put the learners in the same location in a group to meet their requirements of collaborative learning in authentic scenarios. Smart learning environments can provide the most suitable learning path and the most appropriate learning methods for learner.

For the adults and off-campus learners, smart learning environments can organically integrate formal learning with informal learning, to meet the citizen's needs of life-long learning, to adapt to the requirements of campus learning, home learning and social learning so as to truly realize the concept of "seamless learning".

2.4 The Constituent Elements and Technical Features of Smart Learning Environments

As shown in Fig. 2.1, the constituent elements of smart learning environments include six components: resources, tools, learning communities, teaching community, learning ways and teaching ways.

1. Smart learning environments mainly consist of the six elements of learning resources, intelligent tools, learning community, teaching community, learning ways and teaching ways.
2. Learners and teachers interrelate and interact with the other four elements by the teaching ways and learning ways, so as to promote the effective learning of learners together. If the learning ways and teaching ways were taken out, smart learning environments cannot be regarded as learning environments.
3. The occurrence of effective learning is the mutual result of individual knowledge construction and group knowledge construction. Learning community emphasizes the interaction, collaboration and exchange of learners, while teaching community is a continuum where teachers learn together, work collaboratively to pursue continuing professional development.
4. Learning resources and intelligent tools provide support of both learning community and teaching community. The development of learning community and teaching community is inseparable from the mutual effects of resources and

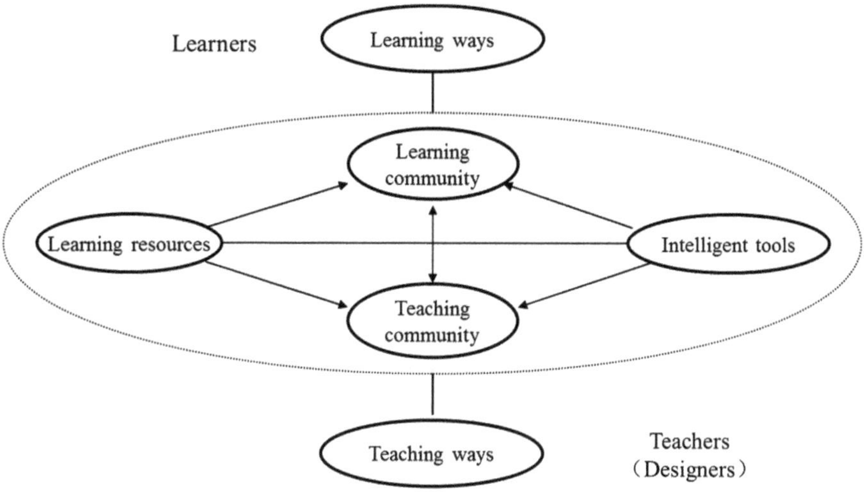

Fig. 2.1 The system model of smart learning environments

tools. All kinds of intelligent tools provide comprehensive support of the "intelligence" of the learning environments. At the same time, learning community and teaching community advance the evolution of resources and tools.

The technical features of smart learning environments are mainly reflected in the five aspects: tracking, recognizing, connecting and awareness, with the aim to promote easy, engaged and effective learning of learners.

1. Tracking learning process. Smart learning environments can be aware of and track the status of learners in the aspects of knowledge acquisition, classroom interaction, group collaboration, etc. By using technologies of motion capture, emotion computing and eye movement, SLE can track the learning process, analyze learning outcomes and establish the learner model, which provides important basis for providing more comprehensive and accurate assessment of the learning effects of the learners.
2. Recognizing learning scenario. Smart learning environments can provide personalized resources and tools for learners according to the learner model and learning scenarios to facilitate the occurrence of effective learning. It can recognize learning scenarios, including learning time, learning place, learning peers and learning activities. The recognition of learning scenarios provides support of teaching activities.
3. Awareness of physical environment. Smart learning environments can monitor air, temperature, light, sound, smell and other physical environmental factors with sensor technology to provide learners with a comfortable physical environment.

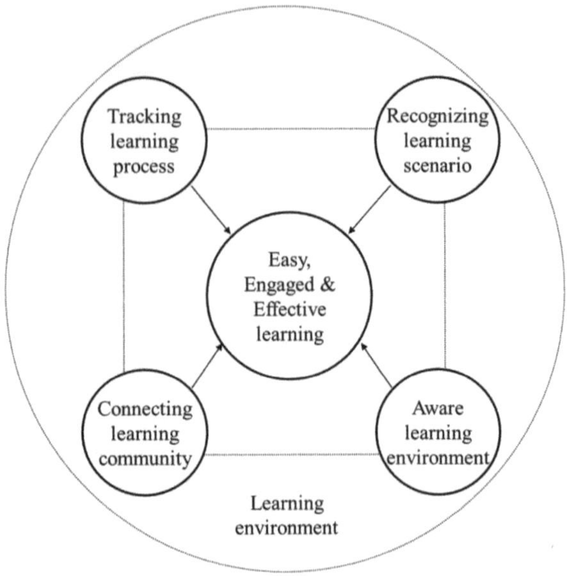

Fig. 2.2 TRACE^3 functional model of smart learning environments

4. Connecting learning community. Smart learning environments can set up learning community for specific learning scenarios and provide support of the learners for their effective connecting and using learning community for communication and exchange.
5. Easy, engaged and effective learning. The objective of smart learning environments is to create the tracking, recognizing, awareness and connecting conditions for learning so as to promote the easy, engaged and effective learning of the learners.

The technical features of smart learning environments, described above have also their functional requirements, referred to as TRACE^3 functional model of smart learning environments shown in Fig. 2.2.

2.5 Typical Smart Learning Environments

With the changes of learning scenarios, the requirements of learners in smart learning environments are different from the previous generation of learners. Huang et al. (2010) believe that learning scenario refers to the comprehensive description of one or a series of learning events or learning activities, that comprehensively describe the four elements of a learning scenario that are: learning time, learning place, learning peer and learning activities. Accordingly, the learning scenarios can be categorized into five typical learning scenarios, which are "classroom learning", "self-learning", "inquiry learning", "learning in doing" and "learning in working".

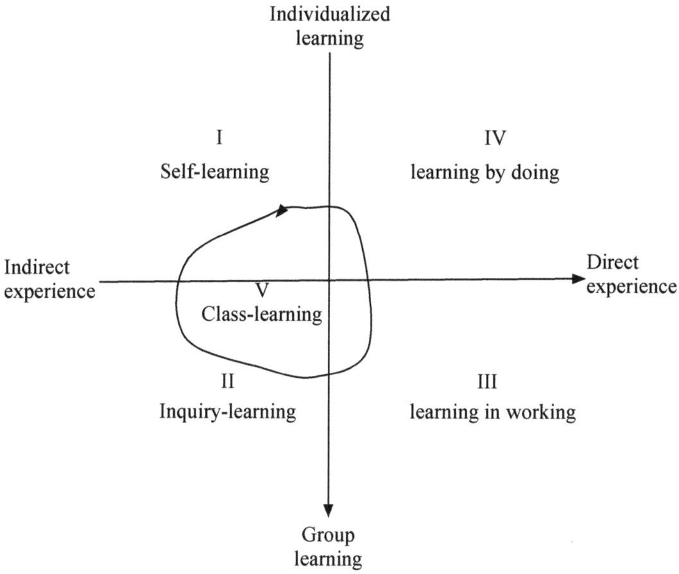

Fig. 2.3 The classification framework of learning scenarios

If the learning scenarios are studied in the two dimensions of the approaches of knowledge construction and those of knowledge acquisition, the classification framework of learning scenarios can be summarized as shown in Fig. 2.3.

Knowledge construction and knowledge acquisition are the two dimensions of studying the diversity of learning scenarios. Each dimension has two different orientations, which are individual learning and community learning of knowledge construction and indirect experience and direct experience of knowledge acquisition. Correspondingly the four quadrants of two-dimensional coordinates exist four learning scenarios, which are "self-learning", "inquiry learning", "learning in doing" and "learning in working".

According to the categories of learning scenarios, smart learning environments should provide the support of "self-learning", "inquiry learning", "learning in doing", "learning in working" and "classroom learning".

2.5.1 Smart Learning Environment Supporting "Self-learning"

"Self-learning" refers to a pre-agreed or voluntary learning behavior of learners, usually without teachers smart learning d mentoring, but with specific learning contents, pre-set learning objectives and evaluation methods. This type of scenarios is more flexible without fixed time and location, but in the learning process the learners are easy to feel lonely and difficult to get help, which has high demands for the readability of the learning materials and personal interests in learning. The smart learning environments that is suitable for this type of learning scenario should have the following features:

1. Learning Resources. In terms of the contents, the provided learning resources include digitalized media materials, online or offline digitalized courses, etc. In terms of the form, the provided learning resources should be of clear structures and lively forms, with the features of self-organization and rich media. In terms of the approach of acquisition, it allows the learners to have free access to the contents that have close relevance to the learning themes, which can be adopted at any time they want.
2. Intelligent tools. In order to accomplish the pre-set learning objectives, the tools typically include learner model measurement tools, information delivery tools, learning track recording tools and learning outcome evaluation tools, etc.
3. Learning community. The knowledge construction of learners is the mutual results of individual construction and community construction. The learning scenario of "self-learning" tends to the individual knowledge construction, while the establishment of learning community aims to encourage regular communication and exchanges among the learners in order to relieve the sense of loneliness in self-learning.
4. Teaching community. In order to promote the self-learning of the learners, teachers establish learning communities to study the teaching contents, learning support approaches, teaching methods, the design of learning resources and learning tools together. The establishment of teaching community contributes to professional development of teachers and also to the smooth progress of instruction.

2.5.2 *Smart Learning Environment Supporting "Inquiry Learning"*

"Inquiry learning" refers to the learning form with participation in small groups and discussion as the main communication approach, which usually have specific discussion topics, moderate scales of members and strong organizers. The participants in this learning scenario can easily become interested, but the organizers are required to have strong organizational skills and good interpersonal relationship. The smart learning environments that is suitable for this learning scenario should have the following features:

1. Learning Resources. This type of learning scenario does not require a large scale of resources. It allows the learners to access to the contents correlated to the learning topics for the use at any time, but requires the teachers or organizers to prepare a high-quality inquiry topic in advance.

2. Intelligent tools. In order to accomplish the pre-set learning objectives, the tools typically include learner presentation tools, learner interaction analysis tools and learning outcome evaluation tools, etc.
3. Learning community. This type of learning scenarios tends to community knowledge construction and the establishment of learning community is the foundation of inquiry learning.
4. Teaching community. The teaching community established in this scenario is to facilitate the communications among teachers of the same subject and promote the development of the instructional design of theme-based classroom teaching.

2.5.3 Smart Learning Environment Supporting of "Learning by Doing"

"Learning by doing" refers to the learning form of implanting the activities of o facilitate the communications among teachers of the same subject and promote the development of the learning tasks and learning objectives, that of evaluation methods and tasks, that of support services and students and that of organizational forms and learning environments. The participants in this type of learning scenario can easily develop interest and acquire better learning results. As it may be difficult to get help when encountering difficulties, they are more dependent on the design of working tasks and the support services of learning. The smart learning environments that are suitable for this learning scenario should have the following features:

1. Learning Resources. In terms of the contents, the provided resources usually include the background information of working tasks and the support services of learning. In term of forms, the provided resources should be of clear structures, lively forms, with the features of self-organization, rich media and ubiquity. In terms of the approach of acquisition, it allows the learners to have free access to the contents that have close relevance to the learning themes, which can be adopted at any time they want.
2. Intelligent tools. In order to accomplish the pre-set learning objectives, the tools typically include learner model measurement tools, information delivery tools, learning track recording tools and learning outcome evaluation tools, etc.
3. Learning community. The knowledge construction of learners is the mutual results of individual construction and community construction. The learning scenario of "learning by doing" tends to promote individual knowledge construction, while the establishment of learning community aims to encourage regular communication and exchanges among the learners.
4. Teaching community. The teaching community in this type of learning scenario can help to predict in advance the difficulties that may take place in the "learning by doing" of the learners and provide relevant support.

2.5.4 *Smart Learning Environment Supporting "Learning by Working"*

The learning scenario of "learning by working" is a learning from experience gathered during in practical work, which is commonly used in corporation training. It usually requires the contents based on practical work, working tasks matching with the intensity of work and interpersonal relationship suitable for learning. Learners normally have interests in learning in working, but they are often unable to handle the relationship between working and learning, so it is of greatly importance to create the learning atmosphere in the organizations and improve the personal learning skills. The smart learning environments that are suitable for this type of learning scenario should have the following features:

1. Learning Resources. The learners usually put forward the practical problems or tasks in their work in the form of topics to cause the mutual discussion among their fellow leaners. The provided resources include digitalized media materials related to the topics, online or offline digitalized courses, etc. In terms of the forms, the provided resources may be loose and unstructured, but have to relate to the topics. In terms of the approach of acquisition, it allows the learners to have free access to the contents that have close relevance to the learning themes, which can be adopted at any time they want.
2. Intelligent tools. In order to accomplish the pre-set learning objectives, the tools typically include learner model measurement tools, information delivery tools, learning track recording tools and learning outcome evaluation tools, etc.
3. Learning community. The knowledge construction of learners is the mutual results of individual construction and community construction. The learning scenario of "learning by working" tends to the individual knowledge construction, while the establishment of learning community aims to encourage regular communication and exchanges among the learners.
4. Teaching community. The teaching community of this type of learning scenario tends to provide support of the difficulties that the learners may have in their "learning by working".

2.5.5 *Smart Learning Environment Supporting "Classroom Learning"*

The learning scenario of "classroom learning" refers to the learning in a real classroom or similar environments, which is a collective learning behavior, usually in the form of classes with fixed teaching environment, the teachers for face-to-face instruction and prepared teaching contents and assessment requirements. This type of scenario mainly focuses on knowledge delivery, which usually lacks of exchange opportunities. The learning outcomes are highly dependent on the teachers' teaching skills and the learners' existing foundations and their interests in the contents. The smart learning environments that are suitable for this type of learning scenario should have the following features:

1. Learning Resources. In terms of learning contents, the provided resources are well-structured media materials designed by the teachers. In terms of content control, it generally does not allow the learners to have free access to or browse the contents that are irrelevant to the learning topics, so the learners can only acquire the contents that have greater relevance to the teaching topics, such as teachers' lecture scripts, special topic websites and special topic resource base and so on.
2. Intelligent tools. In order to accomplish the pre-set learning objectives, the tools typically include learner model measurement tools, information delivery tools, learning track recording tools and learning outcome evaluation tools, etc.
3. Learning communities. The knowledge construction of learners is the mutual results of individual construction and community construction. The learning scenario of "classroom learning" tends to the individual knowledge construction, while the establishment of learning community aims to encourage regular communication and exchanges among the learners.
4. Teaching community. The teaching community of this type of learning scenario tens to establish learning community for the teachers to study the teaching contents, learning support approaches, teaching methods, the design of learning resources and learning tools together. The establishment of teaching community contributes to professional development of teachers and also to the smooth progress of instruction.

2.6 Discussion

The smart learning environments not only consist of intelligent tools, learning resources, learning community and teaching community, but also the learning ways and teaching ways, which make the learning environments "smart" according to the different needs of learners. The TRACE^3 model proposed in the chapter is just a tentative effort to describe the functions of smart learning environment. More research should be done on the issues related to advancing current learning environments towards smart learning environments. These research should investigate such issues like the limitations of existing learning environments, need for reform of educational system, innovative uses of emerging pedagogical approaches and technologies, sharing and promotion of best practices, design and implementation of smart learning environments. We call for action in the following areas for promoting SLEs research and practice:

1. Research on smart learning environments calls for efforts from different areas such as education, computer science, psychology and engineering. The communication and collaboration in these areas are the basis for the development of SLEs.

2. New emerging technologies such as intelligent system, the Internet of Things, cloud computing should be integrated into the smart learning environments to support learning and teaching.
3. Personalized learning is the basic rule for design of smart learning environments, and high users' experience is the first principle. The research on how to improve the users' experience in SLE seems to be especially important.
4. The role of teachers in smart learning environments must change from that of an instructor to that of a facilitator. The community of teachers is a leading element of SLEs. Technological Pedagogical Content Knowledge (TPACK) has emerged as a useful frame for describing and understanding the goals for technology use in faculty development.
5. International collaboration on principles of smart learning and among different researchers is necessary to share case studies, promote innovations and best practices.

References

Chan, T.-W., Roschelle, J., Hsi, S., Kinshuk, Sharples, M., Brown, T., et al. (2006). One-to-one technology-enhanced learning: An opportunity for global research collab-oration. *Research and Practice in Technology Enhanced Learning, 1*(1), 3–29.

Chin, K.W. (1997). *Smart learning environment model for secondary schools in Malaysia: An overview* [*EB/OL*]. 10 October, 2011. http://www.apdip.net/projects/seminars/it-policy/cn/resources/kangwaichin/smartlearning-mimos.ppt

Huang, R., Chen, G., Zhang, J., & Wang, Y. (2010). Research on informationization learning mode and its digital resource form. *Modern Distance Educational Research, 6*, 68–73.

Huang, R., Hu, Y., & Yang, J. (2015). Improving learner experience in the technology rich classrooms. In: Kinshuk & R. Huang (Eds.), *Ubiquitous Learning Environments and Technologies* (pp. 243–258). Berlin, Heidelberg: Springer. doi:10.1007/978-3-662-44659-1.

Hwang, G.-J. (2014). Definition, framework and research issues of smart learning environments—a context—aware ubiquitous learning perspective. *Smart Learning Environments, 1*(1), 4. doi:10.1186/s40561-014-0004-5.

Kinshuk, Chen, N. S., Cheng, I. L., & Chew, S. W. (2016). Evolution is not enough: Revolutionizing current learning environments to smart learning environments. *International Journal of Artificial Intelligence in Education, 26*(2), 561–581. doi:10.1007/s40593-016-0108-x.

Koper, R. (2014). Conditions for effective smart learning environments. *Smart Learning Environments, 1*(1). doi:10.1186/s40561-014-0005-4

Kwok, L. (2015). A vision for the development of i-campus. *Smart Learning Environments, 2*(1), 2. doi:10.1186/s40561-015-0009-8.

Li, B., Kong, S. C., & Chen, G. (2015). Development and validation of the smart classroom inventory. *Smart Learning Environments, 2*(1). doi:10.1186/s40561-015-0012-0.

Long, P. D., & Ehrmann, S. C. (2005). Future of the learning space: Breaking out of the box. *Educause Review*, 40(August), 42–58. Retrieved from http://www.educause.edu/EDUCAUSE+Review/EDUCAUSEReviewMagazineVolume40/FutureoftheLearningSpaceBreaki/157992

Lui, M., & Slotta, J. D. (2014). Immersive simulations for smart classrooms: Exploring evolutionary concepts in secondary science. *Technology, Pedagogy and Education, 23*(1), 57–80. doi:10.1080/1475939X.2013.838452.

Sevindik, T. (2010). Future's learning environments in health education: The effects of smart classrooms on the academic achievements of the students at health college. *Telematics and Informatics, 27*(3), 314–322. doi:10.1016/j.tele.2009.08.001.

Spector, J. M. (2014). Conceptualizing the emerging field of smart learning environments. *Smart Learning Environments, 1*(1). Retrieved from http://www.slejournal.com/content/1/1/2

Yang, J., Yu, H., Gong, C., Kinshuk. (2015). The research focus and future trends of smart learning environments. *E-educational research, 5*, 85–88, 95.

Zhong, G., & Zhang, X. (2006). A Building of the current intelligent learning environment model. *Computer Science, 1*, 170–171.

Chapter 3
Characteristics and Framework of Smart Learning

Abstract Considering few research that has been done on differences between smart learning, digital learning and traditional learning, this chapter aims to identify these differences from the smart city perspective. A framework for smart learning is proposed to show the components of smart learning, that include four levels: learner, supportive technology, core elements of learning scenarios, and logical laws for teaching and learning. We also describe the four principles and four kinds of supportive technologies. At the end of chapter, the relationship between smart learning and smart city is discussed.

Keywords Smart learning · Concept · Framework · Smart city

3.1 Introduction

Learning in smart city happens everywhere, not only as formal learning at schools, but also as informal learning and non-formal learning at workplace society, and homes. Informal learning happens when people are interacting with other people in person or by means of social media It can be a result of any activities during which people need to find information needed to understand something, or just to entertain themselves (Yang et al. 2015).

For example, people go on Facebook or other social media, or people write blogs or read blogs and all those activities create knowledge, however they never think about capturing that information intentionally. When somebody is talking with someone via Skype, that also creates knowledge. Non-formal learning is when learner actually wants to learn something, but does not think he or she needs any formal qualification (Yang et al. 2015). For example, if you want to do some programing in c#, you will just go on the Internet and find some tutorial on C#, or get a book and understand how C# programming is done. Nowadays, MOOCs have emerged that help significantly in this direction.

D. Liu et al., *Smart Learning in Smart Cities*, Lecture Notes
in Educational Technology, DOI 10.1007/978-981-10-4343-7_3

The objective of smart learning environments is to support all kinds of formal learning, non-formal learning and informal learning activities to make learning easy, engaged and effective, or smart. With emerging new technologies that can be used in learning, the concept of smart learning is often appeared in literatures.

Yang et al. (2008) believe that context aware and ubiquitous learning is by nature close the notion of smart learning. Context aware and ubiquitous learning is a computer supported learning paradigm that identifies learners' surrounding context and social situation to provide integrated, interoperable, pervasive, and seamless learning experiences.

Noh (2011) claims that the concept of smart learning emerged as a response to the limitations of e-learning, as well as a result of changes brought to the educational paradigm by advancement in smart devices and technologies. Kim et al. (2011) maintain that the concept of smart learning plays an important role in the creation of an efficient learning environment because it offers personalized contents and can be easily adapted to the current model of education. It also provides learners with rich resources and environment convenient for communication. For workplace learning, Lee et al. (2014) argue that smart learning enables employees' learning to take place anywhere and anytime., and allows to resolve the constraints of online learning, In fact, ubiquitous access is the basic demand for smart learning, not only at workplace, but also in other formal learning settings and informal learning scenarios.

Hwang (2015) mentioned that smart learning was a new concept for developing more powerful and helpful learning environments to provide easy and engaged learning for students, rather than the existing technology-enhanced learning approaches by incorporating new technologies and new criteria for learning. However, being smart does not mean being simply "intelligent" or exclusively "digital", and does not mean either going back to retrace the conceptual path already explored without particular success by the research on Artificial Intelligence and on the Intelligent Tutoring Systems (Giovannella 2014).

The above analysis indicates that smart learning was generally regarded as a new paradigm of learning by using emerging technologies to provide ubiquitous access to learning. However, few publications can be found on the characters of smart learning, and the differences between smart learning, digital learning or traditional learning, especially from the smart city perspective. The following Sect. 3.2 and Sect. 3.3 illustrate the characteristics and framework of smart learning; the relationship between smart learning and smart city is discussed in Sect. 3.4.

3.2 Characteristics of Smart Learning

Learning generally involves a stable and persisting change in what a person or group of people know and can do (Spector 2014). Therefore, smart learning aims at making the stable and persisting change in what a person or group of people know and can do in a smarter way.

There are three words related the concept of smart learning: "Smart", "Intelligent" and "Wisdom". "Smart" means being capable of making adjustments that resemble those resulting from human decisions, chiefly by means of electronic sensors and computer technology. "Intelligent" requires having or showing the ability to easily learn or understand things or to deal with new or difficult situations. "Wisdom" is related to knowledge of what is proper or reasonable, or to a good sense of judgment. The three words implies different aspects of "being smart" and thus shall be reasonably used in suitable, specific situations (Huang 2014).

Smart learning may be seen as a process which cannot be put into application without the support from smart learning environments. Smart learning emphasizes the interaction between learner and the surrounding environment. Smart learning environment is a learning place or an activity space that can be aware of learning scenarios, identify the characteristics of learners, provide appropriate learning resources, convenient interaction tools, automatically record the learning process, and evaluate learning outcomes in order to promote the effective learning of the learners (Huang et al. 2012). Smart learning environments combine the physical and virtual environments to provide individual-earner-specified supports and services. Their technical characteristics are demonstrated in the following four aspects: process recording, context identification, community interconnection, and environmental perception. Spector (2014) maintains that the difference between smart learning environments and ordinary learning environments was that design of the former makes it possible to change the environment based on the variation of learner's comprehensions, intelligence and learning capabilities.

In a smart learning environment, a learner can learn at anytime, anywhere, in any way and at any pace, which is short for "4A". Such an environment is capable of supporting "Easy Learning", "Engaged Learning" and "Effective Learning", which is short for "3E". In such smart learning environment, the easy, engaged, and effective learning (3E) at anytime, anywhere, in any way and at any pace (4A) could be regarded as smart learning. The character of smart learning was shown in Fig. 3.1, which is short for DEEE@4A with D stands for diverse.

The differences of smart learning, traditional classroom learning, and digital learning are shown in Table 3.1. Traditional classroom learning refers to learning in traditional classroom environments with chalks and blackboard as teaching aids;

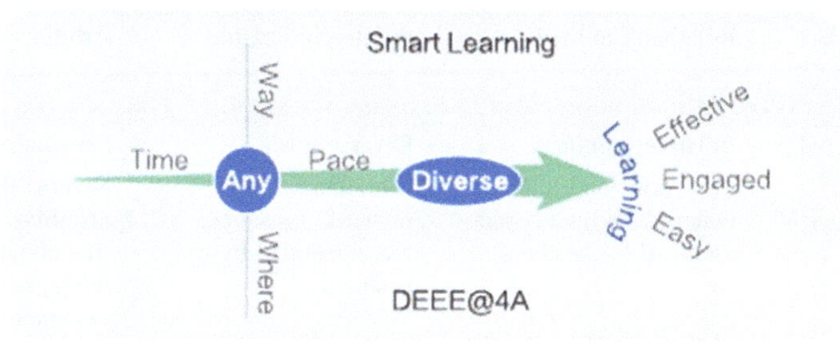

Fig. 3.1 Characteristics of smart learning

digital learning refers to on-line learning environment or classrooms equipped with multiple technologies; smart learning is learning with the character of DEEE@4A in smart learning environments.

Face to face learning in traditional classrooms mainly involves refined knowledge learning processes which stress unified standards, unified learning paces and unified examinations. The learning paths that students take are linear and of the

Table 3.1 Comparison among traditional classroom learning, digital learning and smart learning

	Traditional classroom learning	Digital learning	Smart learning
Learning result and forms	Precision processing of knowledge, In unified forms	Knowledge interconnectivity, in diverse forms	Knowledge interconnectivity, self-adjustment
Learning task	Homogenization	Diversity	Personalization, differentiation
Learning method	Listening and teaching based	Blended learning (listening and teaching + online learning)	Seamless learning
Teaching strategy	Teaching based, but complemented with asking questions and discussion	Application of multiple strategies	Personalized learning instruction
Learning support	Face-to-face Q&A and tutoring	Online communication and support	Multi-channel communication and intelligent system support
Learning assessment	Standardized tests and exams	Online tests at anytime	Adaptable tests
Learning community and ways of participation	Groups and class, School arrangement	Virtual community oriented towards themes, Apply for participating	Virtual community oriented towards themes, Automatic matching and recommendation
Learning space	Fixed physical space	Physical and virtual space	Intelligent learning space
Learning pace and time sequence	Relatively unified	Relatively flexible	Arbitrary pace
Learning goal	Relatively unified	Diverse goals	Personalized goal
Learning resources and source	Textbooks and tutorial materials in hard copies, arranged by teachers	E-textbooks and network resources, recommended by teachers	Diverse digital resources free choice and intelligent recommendation
Learning media	Paper media only	Paper media, internet media	Across terminals, rich media

same nature, which results in monotonous, relatively rigid learning methods which further impedes the cultivation of creative capabilities among students. Learning in smart environments emphasizes the interconnection between knowledge nodes and aims to build a variously-typed, diversified, pluralistically-evaluated teaching and learning ecology. In such a learning method, learners are provided with diversified learning paths such as the linear type, point to surface type or whole to part type paths (Huang 2014).

Smart learning environments provide learners with room for deep-level communication. It can expand innovation spaces, enrich the depth and breadth of learner's knowledge structures, and create diversified supporting environments for them. The new model of instruction based on smart learning environments will incline towards the "4A" model under which smart learning pays more attention to students' diversities and individual variations. This makes it possible to carry out the educational concept of "people first".

Smart learning has three basic characteristics. First is Easy Learning, that is the prerequisite for engaged learning; smart instruction makes the learning process become easy and fun. Second, Engaged Learning, is the precondition for effective learning; only through engaged learning that students are able to perform good communication and cooperation with their fellow students and achieve designed learning objectives. Third, Effective Learning, is the target of smart learning, which means smart learning shall bring the desired learning results effectively.

3.3 Framework of Smart Learning

In order to clearly explain the concept of smart learning we proposed a framework of smart learning, as shown in Fig. 3.2, which takes the learner in the kernel and has four levels. The first level is learning experience, which means that learner can study easily, be engaged and learn effectively (3E) at any time and any place, in any way and at any pace (4A). The second level is created by supporting technologies that include tracking and analysis organization and reconstruction, sensing and adaptation, and assessment and supportive technology. The third level is created by the core elements of learning scenarios, which include learning results, learning tasks, learning methods, learning media, learning resources, learning goals, learning timing, learning space, learning community, learning assessment, learning support, and pedagogical strategies. The fourth level consists of the logical laws for teaching and learning, which refer to the four the basic principles: learning resource matching in learning activities and learning content, teaching logic self-consistency in teaching activity and learning activity, learning experience enrichment in teaching activity and learning space, and timely feedback in learning content and learning space.

From an internal point of view, smart learning framework is built with the learner as the kernel; if the learner performs easily, effectively and is concentrated,

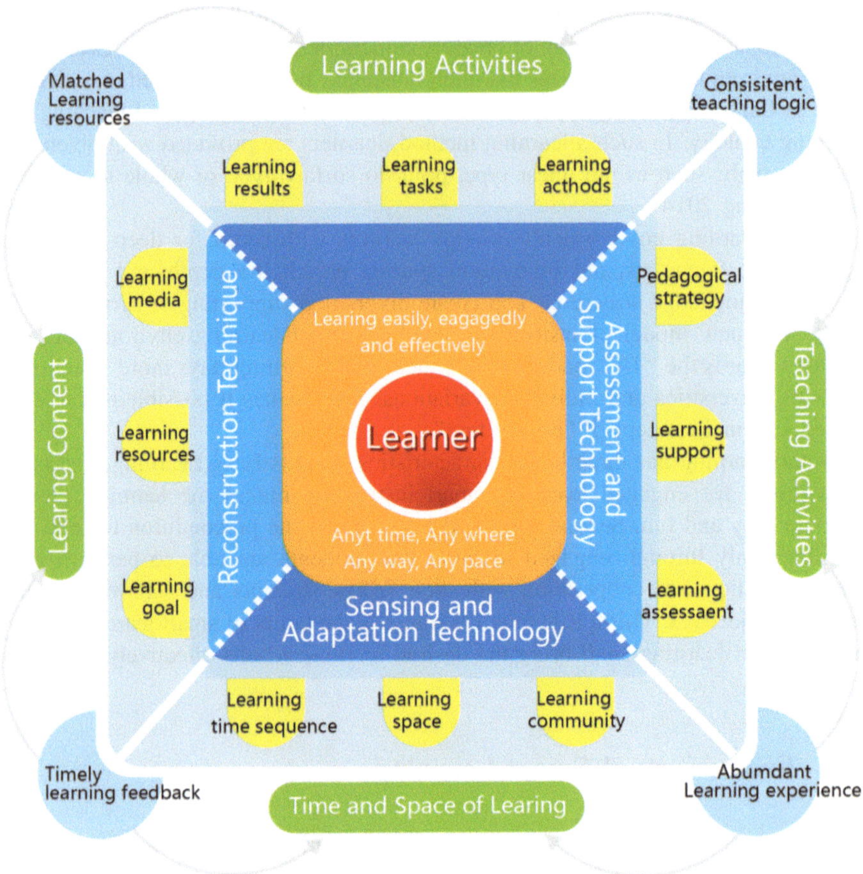

Fig. 3.2 Framework of smart learning

(3E) studies at any time and any place, in any way and at any pace (4A), it means the he/she is a smart learner. From an external point of view, a learning system usually contains four parts: learning activity, teaching activity, learning content, and learning space. Each part is composed of three main elements respectively, and they form the element level. The element level achieves smart learning through the environment provided by the four types of technologies on the technology level. The technology level is a bridge between general learning and smart learning. In the framework of smart learning, the logic of teaching and learning should follow four basic principles of learning resource matching, teaching logic self-consistency, learning experience enrichment, and timely feedback of learning. These are also basis for judging of a smart learning system.

Explanation of relevant concepts

The easiness, absorption and effectiveness of smart learning are all factors taken into consideration by an education provider or education builder. To talk from the support experience level of smart learning environment to learner, the details are as following:

Easiness: To provide fast and convenient learning atmosphere by the surrounding smart learning environment, and make learner's process of completing learning targets become easy.

Engaged: Learning environment can stimulate and facilitate a learner's learning interest, and keep the learner's interest and level of participation at a relatively high state. A learner in such learning environment is absorbed.

Effectiveness: The learning result in the environment of smart learning almost meets or even higher than the expectation.

Explanation of relevant concepts

The explanation of the content of each part in the framework of smart learning is as follows:

Learning Activity: It contains learning task, learning method used and the result of learning after completing the learning task, and these are what will be involved during a learner's learning process.

Teaching Activity: It includes teaching strategies which select, organize, adjust and control students' learning activity, learning support provided during the learning activity, as well as learning assessment during the learning activity and after the learning is completed.

Time and Space of Learning: It includes time sequence of learning for the scheduling of learning resource, learning space for the launching of learning activity (including learning space of physical learning environment and learning space of a virtual learning environment), and social communities of learning for participatory learning and seminar-style of learning.

Learning Content: It contains learning targets directing students learning activity, learning resource during the development of learning activity, resource carrier and learning media.

3.3.1 Four Basic Principles of Smart Learning

In smart learning, teaching and learning should follow four basic principles of learning: resource matching, teaching logic self-consistency, learning experience enrichment, and timely feedback (see Fig. 3.2).

Learning resources should be provided according to learning objectives to support a variety of learning activities, which is called matching of learning resource. Using tracking and analytic technologies to collect data during the learning activities, and to analyze these data to support teaching activities constitutes the teaching logic self-consistency. When students carry out all kinds of learning activities in different time and space, smart learning system provides timely feedback to students by using context awareness and adaptive technologies, as well as assessment and supporting techniques. Students and teachers receive feedback on all aspects of the learning and teaching process, such as learning effects, management efficacy, and teaching benefits.

All kinds of feedback given to students during their learning process should be presented timely, and serve as a guide through the smart learning environment. Emerging learning technologies could provide supports for realizing the four basic principles of smart learning. In the second level of smart learning framework, four types of supportive technologies are included (as shown in Table 3.2):

1. Awareness and adaptive technology: Context awareness technology, and learning adaptive technology by using the basic information of learning time and space in the smart learning framework;
2. Assessment and support technology: Teaching assessment technology and technology supporting by using the basic information of instruction in the smart learning framework;
3. Tracking and analytic technology: Dynamic tracking technology and learning analytic technology by using the basic information of learning activities in the smart learning framework;
4. Organization and reconstruction technology: Knowledge organization and reconstruction technology by using the basic information of learning content in the smart learning framework.

3.3.2 Explicitness of Smart Learning

(1) Explicitness of Learning

Learning in smart city includes two types: the explicit and implicit types. The explicit learning is what people normally think of learning that is formal learning often happens in school. The implicit learning happens at home and in social situations, and people not always think that is learning. Actually, educational types can roughly be categorized into three groups according to their level of explicitness: school education with high explicitness, home education with medium level of explicitness, and social education with low level of explicitness, as shown in Fig. 3.3.

School education with high explicitness level is a learning system with “school” as the environment provided by the government. In school education, people generally would like their kids go to “a good school” “a good class” and get “good grades”. Social education is considered as implicit, which often happen in the learning environment of community learning, enterprise learning, learning in public places. Home is one of the more critical learning places for people; as some citizens

Table 3.2 Four types of support technologies for smart learning

Type	Content	Major technology	Application
Perception and adaptive technology	Context awareness and learning adaptive technology	Artificial intelligence, sensor, auto deduction	To identify the type of learning situation according to information about learning situation: diagnosis of learner's problems and prediction of learner's expectations. This will allow to provide learners with personalized learning resources, find study companions who can mutually cooperate, and accept effective suggestions about the learning activity
	Environment perception technology	RFID, 2-dimensional code, video monitoring	Management to public infrastructures such as Campus Smart Cards, books, instruments, equipment and elevators; intelligent checking-in in classrooms and meetings; switch control; energy-saving control to illumination, air-conditioning and ventilation system control
Assessment and support technology	Teaching assessment technology	Association rules, data mining	To apply association rules to analyze teaching assessment data, discover valuable data pattern, find out relations and rules in-between, play a directive role for the education and teaching activity, and provide decision-making support for teaching administration
	Learning support technology	Augmented reality, 3D printing, rich media, learning terminal	The augmented reality technology and 3D printing technology can enhance users' learning experience. The rich media enhance interactivity, increase audiences' level of participation, improve

(continued)

Table 3.2 (continued)

Type	Content	Major technology	Application
			user's experience, and make the interaction stronger, richer, and more convenient. The personalized learning terminal breaches limitation of the learning place, and makes learning happen not only around the desks
Tracking and analytic technology	Dynamic tracking technology	Motion capture, emotion calculation, eye-movement tracking	Records performance of learners in aspects such as knowledge acquisition, classroom interaction, and group collaboration, tracks the learning process, analyze learning outcome, and builds up learner models
	Learning analytic technology	Classroom teaching effect analysis, interactive text analysis, text mining, video, audio and system log analysis	To provide support for decision-making and optimization in teachers' teaching, and provide effective data support for students' self-guided learning, study crisis pre-warning, and self-assessment. To collect data such as learners' level of study participation, study contents concerned, student and teacher classroom behavior information, learning status and learning resource utilization information. To excavate all kinds of information that learners search from learning resource library and learner information
Organization and reconstruction technology	Organization technology	Learning object, semantic web	To make learning resources easy for learners to retrieve and categorize the resource, increase recall ratio and accuracy of the retrieval, make information more

(continued)

Table 3.2 (continued)

Type	Content	Major technology	Application
			easily be found and integrated by the automatic data mining tools, realize flexible resource sharing, coupling and reusing; at the same time, possesses good expansibility and can be the base for intelligent resource retrieval and pushing, and increase adaptability of the system and personalized service ability to the users
	Reconstruction technique	Learning object, semantic web	Based on different learning demand, through changing structure and way of presentation of content itself, match with detail learning demand

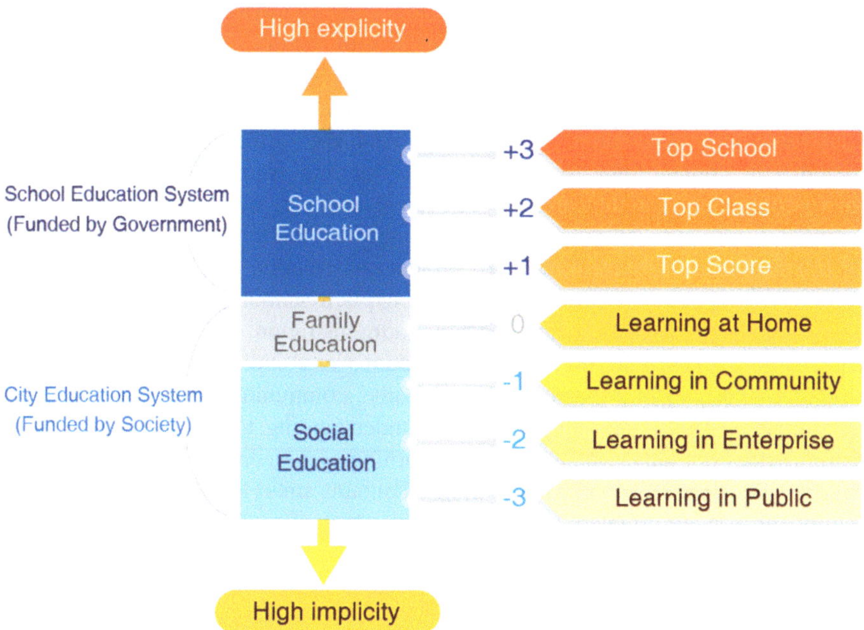

Fig. 3.3 Cognitive explicitness level of learning

have realized the importance of home education to the growth of children and youth, whereas some have not or their level of realization is relatively low, home education is at the middle. It lies between school education with high explicitness level and social education with high implicitness level. There is also a viewpoint which considers home education as one part of the social education.

To mark with the number of levels, learning system can be roughly divided into "elite school" (+3), "experimental class" (+2), "ranking" (+1), home learning (0), community learning (−1), enterprise learning (−2), and public learning (−3) according to the order from high explicitness to high implicitness.

Considering the background of smart city construction, and the level of cognitive explicitness of learning, we could deduce the correspondent relationship between education types and smart learning systems. Within the scope of school education provided by the government, a smart learning system is the high-end form of informatization of school education. Within the scope of family education and social education as provided by the society, a smart learning system is the smart learning environment of a city.

(2) Explicitness of smart learning

Looking at the level of explicit smart learning in Smart City, certain correspondent relations can be built up based on the education type and smart learning system. Smart learning system includes two aspects: school smart learning system and social smart learning system. The school smart learning system can have four parts of contents ranging from macroscopic view to microscopic view, and they are: smart education administering, smart school district (group), smart campus, and smart classroom. And the latter three correspond to "People-to-People Communication", "School-to-School Communication", and "Classroom-to-Classroom Communication" respectively. And the same, social smart learning system can contain four parts of contents ranging from macroscopic view to microscopic view, and they are: smart public learning environment, enterprise smart learning, community smart learning, and home smart learning, as shown in Fig. 3.4.

In smart learning system, the past project construction of educational informatization mostly focus on the investment of explicit school smart learning system, concentrated on education and teaching administration, school informatization construction, classroom digitized construction, and individual web-space construction, and ignores the construction to home, community, enterprise and public environment. This is also the current misunderstanding to smart learning. Smart learning does not only contain the smart learning system embodied as the high-end form of education informatization, but also contains smart learning systems that are family-and-society-oriented, which is also the key of smart learning construction under the concept of Smart City. Enterprise smart learning is an important constituent part in the social smart learning system, which is also one of the indispensable core contents in the construction of smart learning system. To speak of the level of learning, the courses students receive at school are explicit learning, social smart learning is implicit learning, and the latter is the learning method that is more easily ignored. The report of the 16th party congress of the Chinese Communist

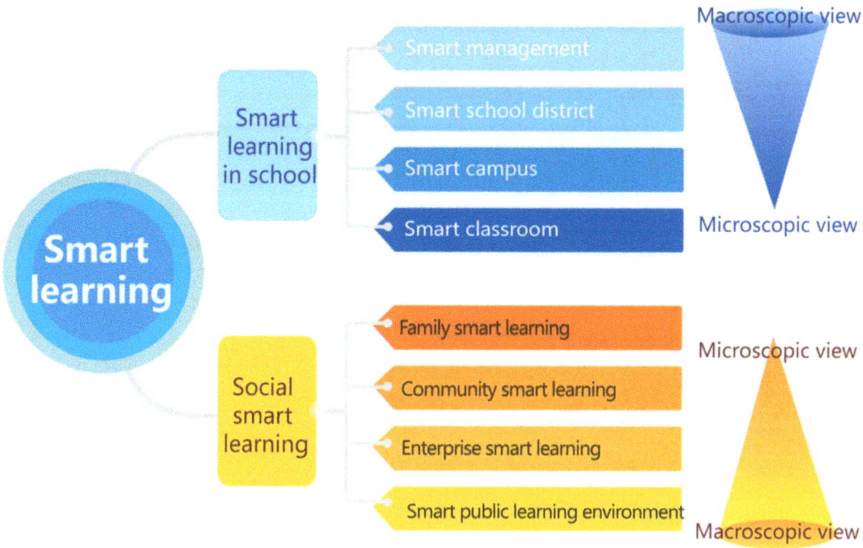

Fig. 3.4 Explicit level of smart learning

Party clearly stated to "form a learning society in which all people learn and pursue life-long learning, which in turn facilitates people's all-round development." Therefore, it is an important path to realize the visualization of learning society to build social smart learning platforms, to highlight the role of implicit learning, and to perfect the function of enterprise learning and home learning.

3.4 Smart Learning and Smart City

3.4.1 Smart Education in the Context of Smart City

Smart education is one of the targets of smart city development and also guarantees the healthy and successful development of a smart city. As the construction of a smart city needs lots of high quality intellectuals, education should be able to provide and cultivate innovative intellectuals.

Smart education can be seen as a kind of system: an educational system that is developed by a school, an area or a country to realize high learning experiences, high content suitability and high teaching efficiency. By taking advantage of modern science and technologies, it can offer a series of differentiated support and on-demand services for students, teachers and parents. Bedsides, the system is able to promote education equity and learning performance by gathering and analyzing the big data of the learner's learning process and learning activities.

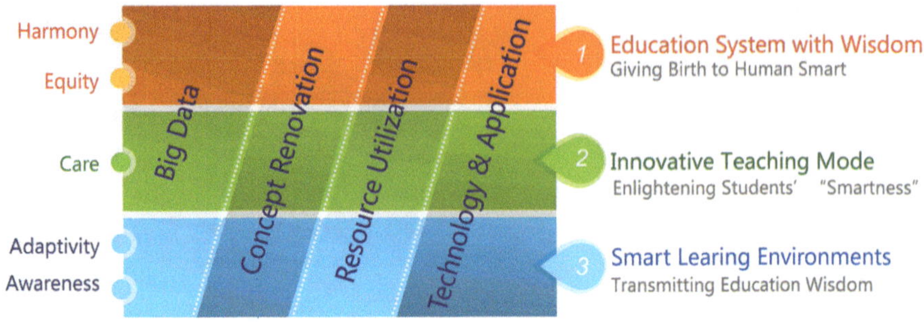

Fig. 3.5 Smart education system

Awareness of the learning environment, adaptivity of the learning content, respect and care to students, education equity among the groups of people, organic integration and harmony of the education elements comprise the intrinsic characteristics of smart education. Specifically, awareness of the learning environment and adaptivity of the learning content fall within the scope of smart learning environments, the function of which is to deliver the "wisdom" of an educational system. The new type of teaching model pays attention to the differences and variation of the students, with the aim to promote easy, engaged and effective learning. Finally, the application of big data methods to analyze and dynamically simulate the subsystems (school arrangement, education finance, channels of employment, student enrollment and their evolution, etc.) can serve as resources for reform schemes and decision-making bases of national education system, school management systems and teaching. The ultimate target is to form modern education system so that it is possible to develop human "wisdom" through developing comprehensive innovative talents and cultivation systems and by promoting education equity between the management zones, between the urban and rural areas and between schools (Huang 2014).

The features of smart education were shown in Fig. 3.5, which includes three levels. The first and fundamental level is smart learning environments with the basic features of sensibility and adaptivity to transmit smartness; The second level is innovative instructional modes with the features of careness to enlighten learner's wisdom; the third level is modern educational system with the features of harmony and equity to develop hum wisdom.

3.4.2 Smart Learning in the Context of Smart City

The key dimensions for the dynamic operation of a city include the supporting systems, learning systems and operating systems. Among these elements, the learning system plays an important role to connect various operating systems, and it is also a kind of social service that serves the people's needs by meeting expectations at the individual and the collective levels. Enterprise's learning system may serve as a good example. It is available in the operating systems of various industries to satisfy the learning needs of individuals and groups of people. The education system can also be seen as an operating system, like the traffic system or financial system, which can undergo informationized construction. The learning system can run through the education system, and become a basic supporting service that can satisfy the educational needs and expectations of individuals.

Based on this, we can define the smart learning system in smart city as a system that can provide comprehensive social services. On the one hand, it is a specific smart learning sector for a smart city, creating a city's smart education system and citizens' learning environments On the other hand, it is also the management element of a smart city, which offers self-organized learning environments. Therefore, it is possible for the smart learning system to connect various operating systems of a smart city. Meanwhile, the smart learning system is capable of comprehensively improving the innovation vitality for the two cores of smart city (vitality for the urban innovation and civil livable experiences).

Explanation of relevant concepts

Creative capital: Creative capital refers to the summation of physical capital, human capital, natural resources and technical knowledge that is capable of providing creativity. In the smart city context, creative capital refers to the element presented in the form of human capital and technical knowledge, mainly including technical reserve, top-level talents and inclusive atmosphere.

Innovative environment: Innovative environment refers to the summation of various factors that influence an innovator's creating activities. Innovative environment in the context of smart city includes convenient traffic, innovative culture and cooperation mechanisms.

The function of smart learning in smart city is shown in Fig. 3.6. Based on the dual-core framework of smart city proposed in Chap. 1, "civil livable experiences" and "vitality for urban innovation" are the two cores of a smart city. On the left of Fig. 3.6, education system and civil learning environment is closely related civil livable experience which include smart learning, smart traffic and smart living. Smart learning is the basic foundation, as it provides top-level talents, technical repository, and inclusive environments which give creative capital opportunities to grow quickly to lead the development of the city.

On the right of Fig. 3.6, urban self-organized learning environments is closely related to "vitality for urban innovation" which include smart economy, smart environment and smart governance. Smart learning is also the basic foundation, as it provides innovative culture, mechanism of collaborative innovation, and convenient traffic which give birth the innovative environment for the self-development of a smart city. The urban self-organized learning environments interact with the

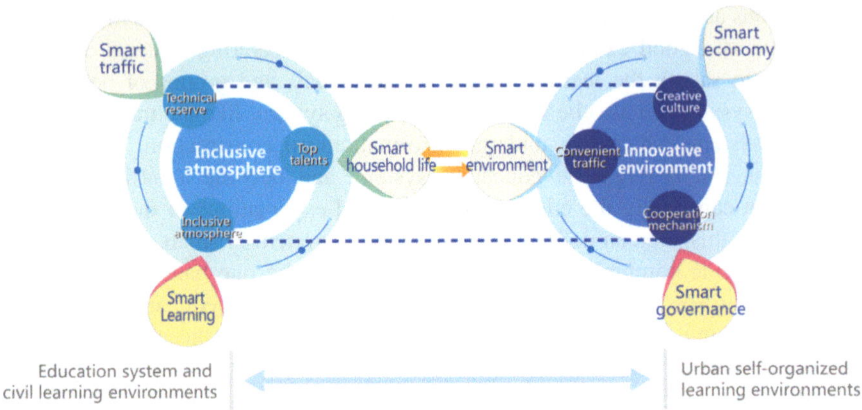

Fig. 3.6 Function of smart learning in smart city

education system and civil learning environments to provide bountiful livable experience and innovative vitality.

In a smart city, the basic output undertaken by smart learning includes three aspects: learning to learn, learning to apply, and learning to self-actualize. Smart learning in such context does not only cover all the contents of smart education occurring in schools, but also includes learning environment construction aimed at providing vocational learning for employees and citizens. Smart learning is of key value to assist the improvement of vitality for urban innovation construction.

On the one hand, the function of smart learning in smart city is to reform school education by applying emerging information technologies to improve the efficiency of education system for cultivating intellectuals that meet the needs of this information society. On the other hand, smart learning in smart city aims to inspire citizen to develop lifelong learning habits of continuous acquisition of knowledge to fit themselves with the fast paces of society development.

From the macro urban city perspective, smart learning is a type of learning provided by a smart city for the purpose of achieving high learning experiences, high content suitability and high learning performance. Smart learning is contained in various city education systems like school education, family education, community education as well as science and culture venues.

3.4.3 Understanding Smart Learning in City Environment

With the wave surging of Internet+, industrialization 4.0, innovation and entrepreneurship, smart campus, learning society, smart city and so on has become hot topic of research and development. How to build a smart campus by school? What is the developmental direction of smart learning environment? What is the smart learning environment in the smart city? All these research questions have always been discussed by many educators, educational leaders and supporters. In order to solve these problems, we should understand smart learning in the city environment.

3.4.3.1 Learning Is More Than a Classroom

Learning environment is the space where teaching and learning activities are conducted. The evolution process of learning environment is from the initial nature forest to schools, private schools and academy private schools and Imperial College, Academy of officer. The learning environment with modern significance was appeared after the 'classroom teaching' put forward by Comenius. Therefore, nowadays, it is generally believed that the learning environment is the classroom and the external environment which influences the learners' learning. From the perspective of educational psychology, learning environment provides external conditions that enable learners to construct knowledge and skills proactively. Therefore, the learning environment can be a classroom, an online classroom, or family classroom, etc. It can be both physical and virtual. As long as it is the external condition that can enables the learners to acquire knowledge and skills, it can be called learning environment.

3.4.3.2 Education Is More Than a School

Educational types can roughly be categorized into three levels according to their level from high explicitness to high implicitness: school education, home education, and social education. School education at high explicitness level is a learning system with "school" as the environment which is provided by the government, and it can be categorized according to the level of explicitness from high to low: "a good school", "a good class", "good grades". Social education at high implicitness level is a learning system with "city" as the environment which is provided by the society, and it can be categorized according to the level of implicitness from high to low (the level of explicitness from low to high): community learning, enterprise learning, public learning. Home learning lies between school education with high explicitness level and social education with high implicitness level. To mark with the number of levels, learning system can be roughly divided into "a good school", "a good class", "good grades", home learning, community learning, enterprise learning, and public learning according to the order from high explicitness to high implicitness.

3.4.3.3 Smart Learning Environment Is More Than an Industry in Smart City

Famous management expert Russell L. Ackoff maintained that the key dimensions of a city's dynamic operation included support systems, learning systems, and business systems. Among them, learning system is a comprehensive support through a variety of business systems, and is supposed to meet their needs and expectations both the individual citizens and organizations. Therefore, we say, smart learning environment is an integrated support system in the smart city system. The basic output of the smart learning environment design includes three aspects: learning to Learn (Formal education), learn to use (Professional education), learn self-realization (Cultural Education). Designing of a smart learning environment should includes not only all parts of the existing smart education program, it should also includes the professional education for the service personnel and creating of the cultural and educational environment for the citizens. Smart learning environment design in the smart city should also support all kinds of education in a city, such as formal education, vocational education and training, social and cultural education.

References

Giovannella, C. (2014). Smart learning eco-system: "FASHION" or "BEEF"? *Journal of E-Learning and Knowledge Society, 10*(3), 15–23.

Huang, R. (2014). Three realms of smart education: Smart learning environment, ICT teaching model and modern educational system. *Modern Distance Education Research, 6,* 3–10.

Huang, R., Yang, J., & Hu, Y. (2012). From digitalized environment to smart environment: The reform and trends of learning environment. *Open Education Research, 18*(1), 75–84.

Hwang, G.-J. (2015). Definition, framework and research issues of smart learning environments—a context-aware ubiquitous learning perspective. *Smart Learning Environments, 1*(1), 1–14. doi:10.1186/s40561-014-0004-5

Kim, S., Song, S., & Yoon, Y. (2011). Smart learning services based on smart cloud computing. *Sensors, 11,* 7835–7850. doi:10.3390/s110807835

Lee, J., Zo, H., & Lee, H. (2014). Smart learning adoption in employees and HRD managers. *British Journal of Educational Technology, 45*(6), 1082–1096. doi:10.1111/bjet.12210

Noh, K. (2011). Smart learning and future education. *Education Information Wednesday Forum,* KERIS. 4 May, 2011.

Spector, J. M. (2014). Conceptualizing the emerging field of smart learning environments. *Smart Learning Environments, 1*(1), 2. doi:10.1186/s40561-014-0002-7

Yang J., Gong C., Yu H., Kinshuk (2015). The research focus and development trend of smart learning environments. *E-educational Research, 5,* 85–88, 95.

Yang, S. J. H., Okamoto, T., & Tseng, S. S. (2008). Context-aware and ubiquitous learning (Guest Editorial). *Educational Technology & Society, 11*(2), 1–2.

Part II
Smart Learning in Specific Domains

Chapter 4
Smart Learning in Digital Campus

Abstract Smart learning can happen anywhere, however, the formal learning in schools and universities is still the basic environment. Digital campus plays an important role in expanding of the school's campus, enriching campus culture, and optimizing teaching, research, management, and service. This chapter investigates the current status of digital campus in China, mainly from the perspective of network construction, setting digital learning terminals, providing learning support systems, and information safety. Desk research method was used in this study, and data about the current status of digital campus from the Statistical Education 2013 to 2014 in China were analyzed. Results reveal that the network access has improved, the learning terminals are enriched, the construction of teaching support service system is unbalanced in different schools, and information safety has been enhanced. Following the analysis of the status, the future trends of digital campus are discussed, and 10 best practices of digital campus in China are introduced.

Keywords Digital campus · Smart learning · Digital learning · Survey

4.1 Introduction

As discussed in the previous chapters, smart learning in smart city includes lots of learning scenarios, both formal learning in school, and informal learning at home and society. In the following chapters we are going to discuss smart learning in different settings, such as schools, classrooms, enterprises, and industries.

The concept of Digital campus is widely used in China's governmental policy documents and academic literature. Digital campus usually refers to the digital environment that integrates a variety of systems for teaching and learning, with the purpose to expand the school's campus, enrich campus culture, and optimize processes such as teaching, research, management, and service Huang et al.(2012a, b).

The development of digital campus in China originated from colleges and universities. In 1994, benefited from the construction of CERNET (China Education and Research Network), construction of campus networks in colleges

D. Liu et al., *Smart Learning in Smart Cities*, Lecture Notes
in Educational Technology, DOI 10.1007/978-981-10-4343-7_4

and universities entered a stage of a high-speed development. The universal construction of digital campuses in basic education started since the initiation of the project "School-to-School connection" in 2000.

In June 2012, the Ministry of Education of China released the "12th Five-Year Plan for the Development of Education in China", and raised the idea of incorporating informatization[1] of education into the rank of national development strategy. On September 5, 2012, during the television conference of national education informatization, vice-prime minister Liu, Yandong (State Councilor at that time) pointed out, that during the 12th Five-Year period, China should concentrate on the construction of "Three Connections and Two Platforms". This means that "Every school can access the broadband network, every class can access high quality resources, and everybody can access cyber learning space", and that China should "build a public service platform of education resources and a public service platform for education administration".

There are two aspects of substantial contents for "Every school can access broadband network". One is to provide broadband access to all kinds of schools of all levels. The other is to incorporate relevant networks in construction of the basic teaching environment in of all kinds of schools of all levels Under the pressure of the project of "Three Connection and Two Platforms", the digital construction of digital campuses campus in primary and secondary schools develops vigorously in the whole country.

The development of digital campus in China should follow four stages:

1. Building campus network environment that implement simple information query services, with relatively smaller amount of application systems and lower interconnectivity and intercommunication.
2. Creating relatively more digital resources; the application systems can be integrated, and provide support for teaching, teaching-research, administration and service.
3. Creating rich digital resources and strong integration of application systems. The facilities supporting software for digital campus should have openness and extendibility, and be able to support teaching, teaching-research, administration and service to the high extent.
4. At this stage digital campus should effectively support teaching and learning, enrich school's campus culture, and truly expand school's time and space dimensions (Huang 2009).

Digital campus can be evaluated on five dimensions: as a network facility, environmental equipment, campus website, resource system, as well as the business system. The performance of the corresponding dimensions for the above four stages is shown in Fig. 4.1.

[1]Informatization is a well-known Chinese academic construct, but is relatively unknown in Western academia. Informatization is the process by which computer information processing technology and transmission means are widely used in all sectors of society.

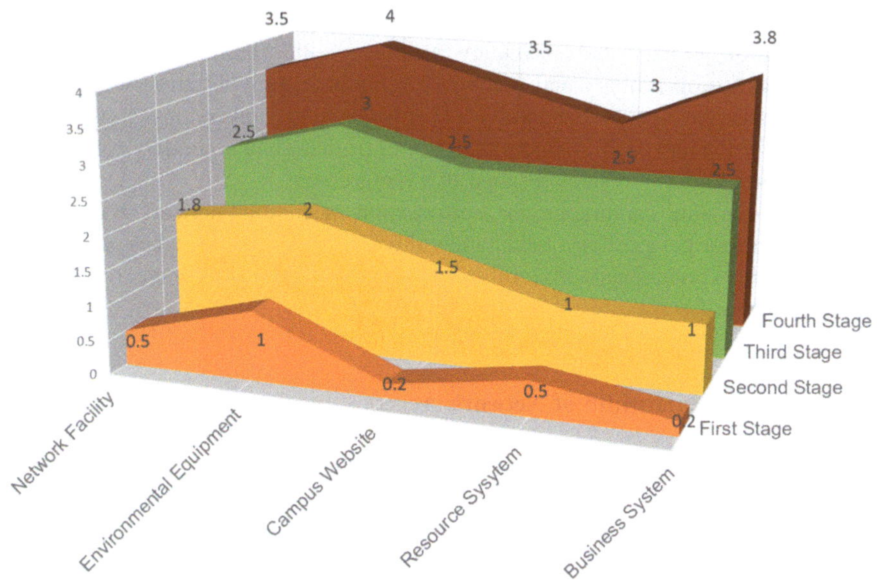

Fig. 4.1 Development stages of digital campus

In constructing of a digital campus, we need to build teaching and learning terminals, provide teaching support service system, and create network's security system. In order to understand the status of digital campus in China, we did a survey by using the desk research method. In this study we used information the data about education from the 2013 and 2014 Statistical Education Data (MoE 2014) released by the Development and Planning Division of the Ministry of Education of the People's Republic of China. The data on educational technology used in this desk research are also from *Development Report about Chinese Education Informatization* (2014) by Wu (2015). We analyzed the overall status of network, digital learning terminals, learning support systems, and network's security systems of institutions in basic education, vocational education, and higher education. The results are presented in the next sections.

4.2 The Current Status of Digital Campus in China

4.2.1 The Condition of Network Access Have Been Improved

Building the on-campus network is an important element in the construction of infrastructure for education informatization. The *Ten-Year Development Plan of Education Informatization (2011–2020)* clearly pointed out, "By the year 2015, broadband network would cover all kinds of schools at all levels; the access bandwidth of primary and secondary schools would reach 100 Mbps, access bandwidth of rural primary and secondary schools at outlying areas would reach 2 Mbps, and the access bandwidth of colleges and universities would reach 1 Gbps." By the end of 2014, more than 6% of primary and secondary schools and 70% colleges and universities reached this requirement.

In the following paragraphs, the construction of campus network will be presented from four perspectives: total bandwidth of network access, types of outgoing networks, backbone bandwidth of campus network, and coverage of wireless campus network.

4.2.1.1 Total Bandwidth of Network Access

In the domain of basic education, by the end of 2014, there had been 6.32% school networks of primary and secondary schools around the country whose total bandwidth reached 100 Mbps above. There were 31.59% schools of which total access bandwidth was between 10 and 100 Mbps, and 31.12% schools of which total bandwidth was 4 Mbps and below, as shown in Fig. 4.2.

Both in 2013 and 2014 the construction of campus networks providing access to the Internet at high school level was better than for the primary schools and junior schools around the nation (Fig. 4.3). In 2014, the proportion of junior schools that built campus network increased apparently in comparison with the year 2013. The number of primary schools that built campus network and the number of schools that had Internet access both had increase largely. However, the proportions of schools that set up campus network with Internet access were still only 39.10 and 77.43% in 2014, and the gaps were still bigger in comparison with those of the junior school and high schools.

The status of bandwidth of network access in vocational institutions in 2014 was shown in Fig. 4.4. The total bandwidth of the network access in secondary vocational schools was at 100M on the average. The total bandwidth of the network access of vocational institutions was at 1000M on average.

The average value of total bandwidth of Internet access of colleges and universities in 2014 was 3800 Mbps. As shown in Fig. 4.5, more than 70% colleges and universities had their access bandwidth reaching or exceeding 1000 Mbps, and reached the requirement for the goal of 2015 according to the *Ten-Year Development Plan for Education Informatization (2011–2020)*, while 22.79% colleges and universities had their Internet access bandwidth reaching 5000 Mbps above.

Among all kinds of colleges and universities, the bandwidth level of Internet access for colleges and universities in 985 Project and 211 Project[2] was far higher

[2]Project 211 is the Chinese government's endeavor aimed at strengthening about 100 institutions of higher education and key disciplinary areas as a notional priority for the 21st century. Project 985 is a constructive project for founding world-class universities in the 21st century conducted by

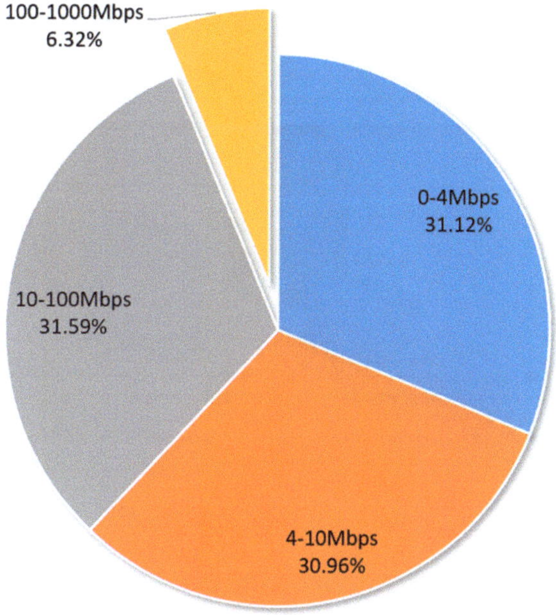

Fig. 4.2 Distribution of total bandwidth of internet access for K-12 Schools around the country

than the other ordinary colleges and universities (see Fig. 4.6). The proportions of access bandwidth above 3000 Mbps for 985 Project and 211 Project at colleges and universities were 88.89% and 59.26% respectively. However, there were still 21.79% of ordinary colleges and universities whose access bandwidth was lower than 500 Mbps.

4.2.1.2 Types of Outgoing Networks

The outgoing methods of broadband network used by primary and secondary schools, vocational institutions, and colleges and universities mainly are China Unicom, China Telecom, China Mobile, metropolitan-area networks (MAN) for regional education, educational satellite, China Education and Research Network (CERNET) etc. The outgoing network types of all kinds of schools are shown in Fig. 4.7. The network of China Telecom is the major outgoing network type used by K-12 schools, vocational education institutions and higher education institutions. Besides, some vocational education and higher education institutions also use CERNET as their outgoing network, but none of the K-12 schools use it. Some

(Footnote 2 continued)

government of People's Republic of China from 1998. Referred from http://www.chinaeducenter.com/en/cedu/ceduproject211.php.

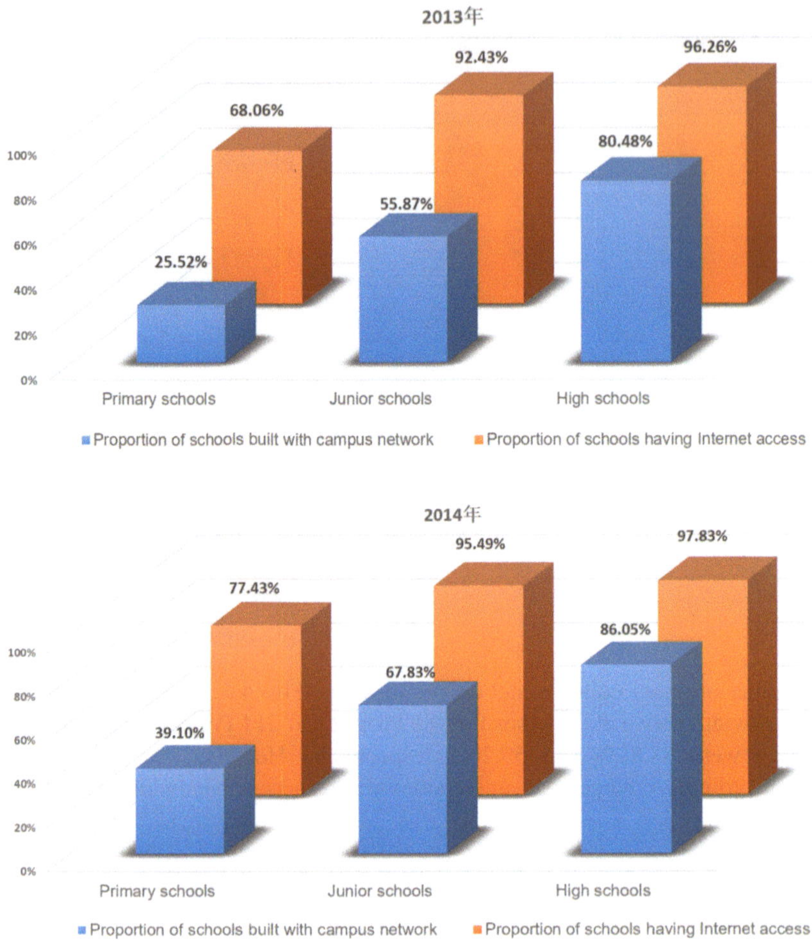

Fig. 4.3 Campus network construction and internet access status of the schools of basic education level around the country in 2013 and 2014

K-12 schools choose educational satellite as their outgoing network, but none of the vocational education and higher education institutions use it.

4.2.1.3 Backbone Bandwidth of Campus Network

The rough distribution of the backbone bandwidth at all kinds of campuses networks is shown in Fig. 4.8. More than 95% colleges and universities nationwide had their bandwidth of backbone campus network reaching 1000M; in which, half of them reached 10,000M. There was quite big difference between secondary

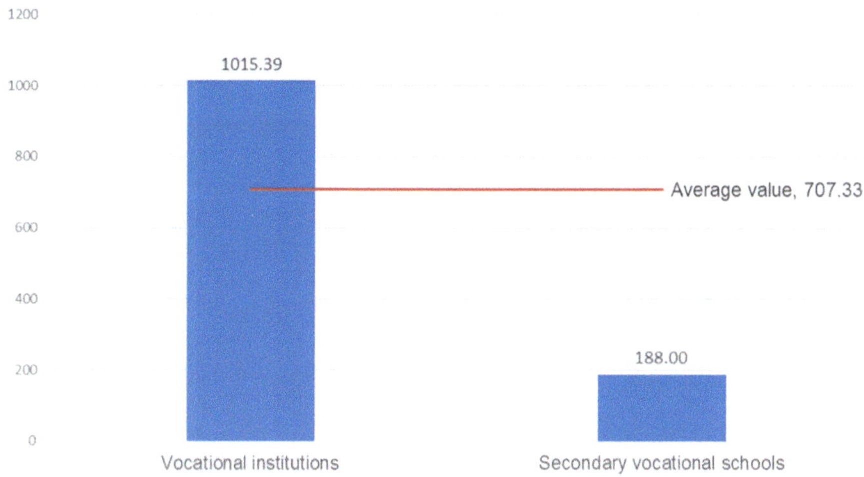

Fig. 4.4 Average total bandwidth of network access for vocational institutions (*unit* Mbps)

Fig. 4.5 Internet access bandwidth for colleges and universities

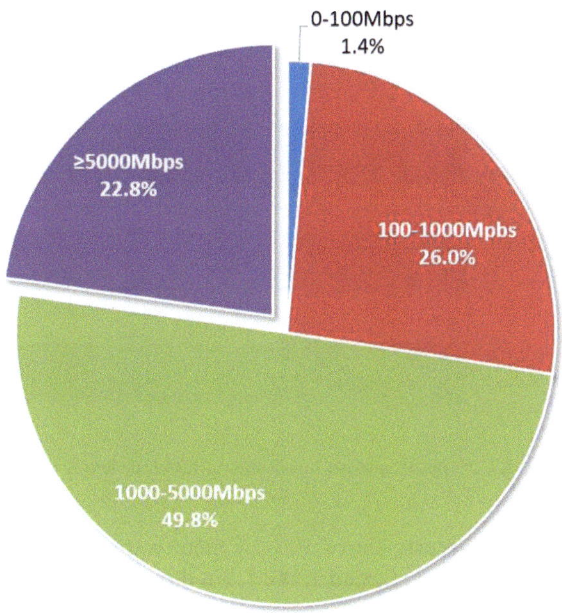

vocational schools and vocational colleges. There were more vocational colleges with 1000M level and 10,000M level, whereas secondary vocational schools had more with 100M level and 1000M level. Among K-12 schools which had had

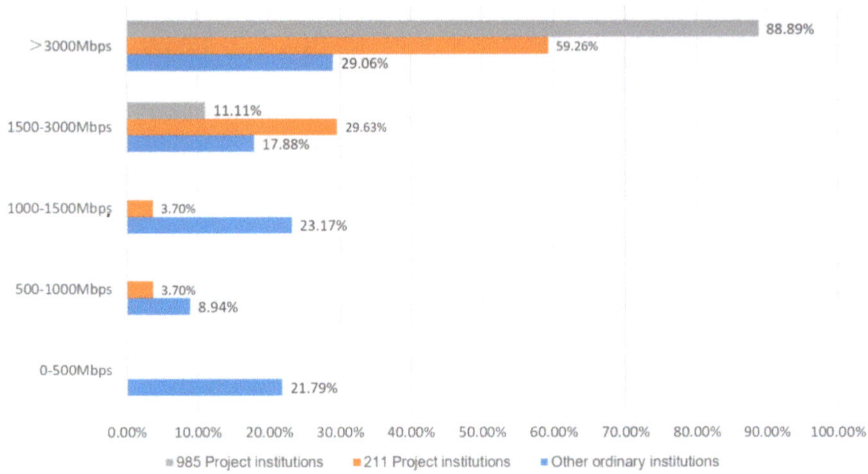

Fig. 4.6 Internet access bandwidth of colleges and universities

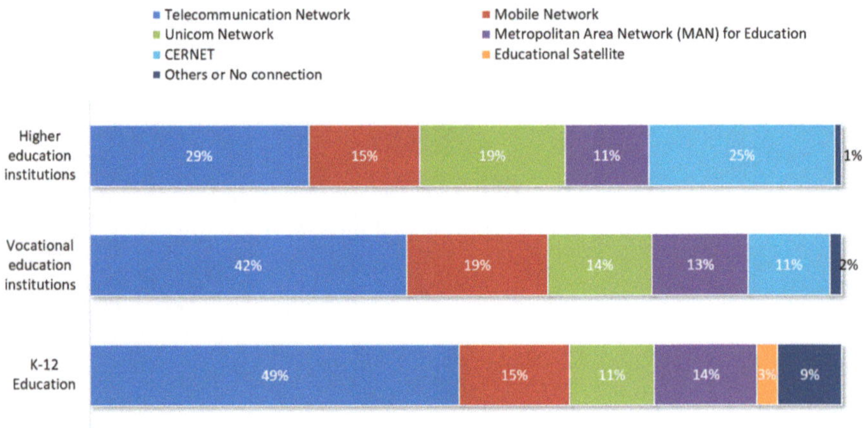

Fig. 4.7 Type of network used among different levels of schools

Internet access, above 50% of them were at 10M level and 100M level, and not many of them reached 1000M, but there were still 33% schools which had not had a campus network yet.[3]

[3]Note: Here the data of backbone bandwidth of campus network for K-12 schools were taken by the end of 2012.

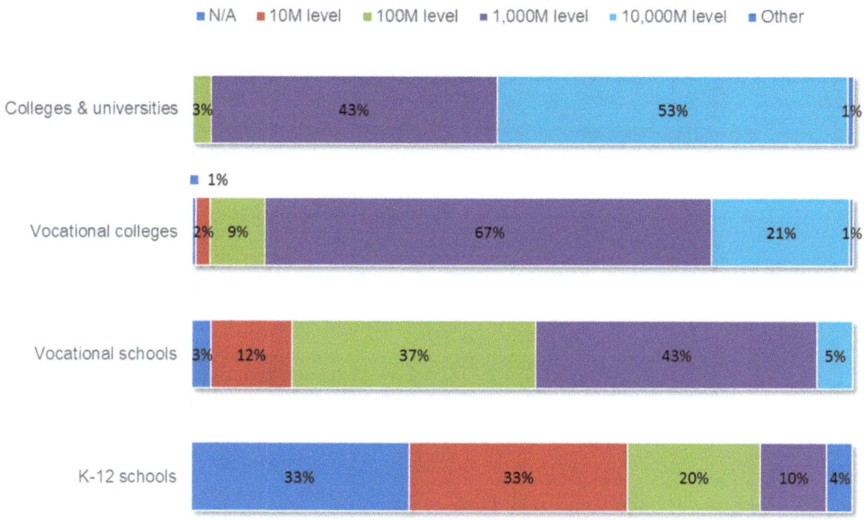

Fig. 4.8 Backbone bandwidth distribution for all kinds of campus networks

4.2.1.4 Wireless Network Coverage on Campus

By the end of 2014, the coverage status of wireless network at all kinds of schools is shown in Fig. 4.9. Among all kinds of schools participating in the investigation, 20% of higher education institutions implemented full wireless coverage, 9% K-12 schools had full wireless coverage, but there was no vocational institution which had full wireless coverage yet. Besides, among K-12 schools, vocational education institutions, and higher education institutions, the ratio of institutions with "20% coverage & below" were all exceeded 40%, as shown in Fig. 4.9.

4.2.2 Learning Terminals Have Been Enriched

Digital learning terminals are an important part in the construction of digital campus. At the current time, digital learning terminals mainly contain desktop computers, notebook computers, tablet computers, intelligent, and cellphones. Below we present results of the survey on different kinds of digital learning terminals in K-12 schools, vocational education and higher education institutions respectively.

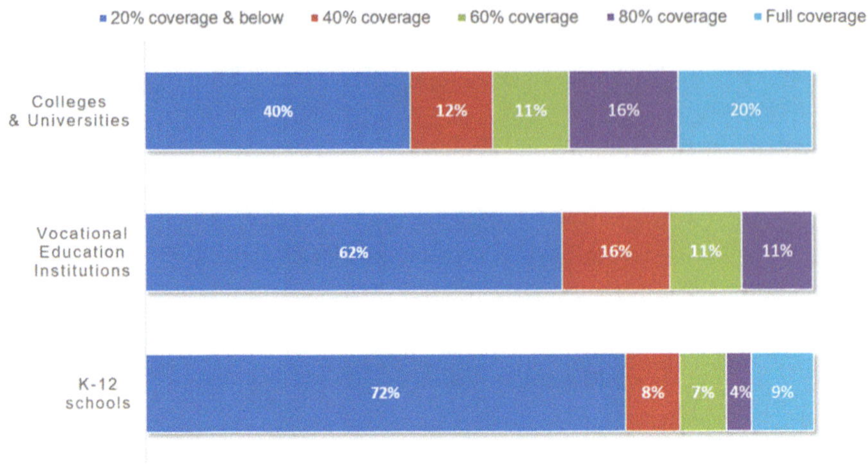

Fig. 4.9 Wireless coverage status of all kinds of schools

4.2.2.1 Digital Learning Terminals in Basic Education

1. **Number of instructional computers in different areas at basic education level**

By the end of 2013, instructional computer number at basic education level in different areas was shown in Fig. 4.10, and the number of tablet computers for instructional use was shown in Fig. 4.11.

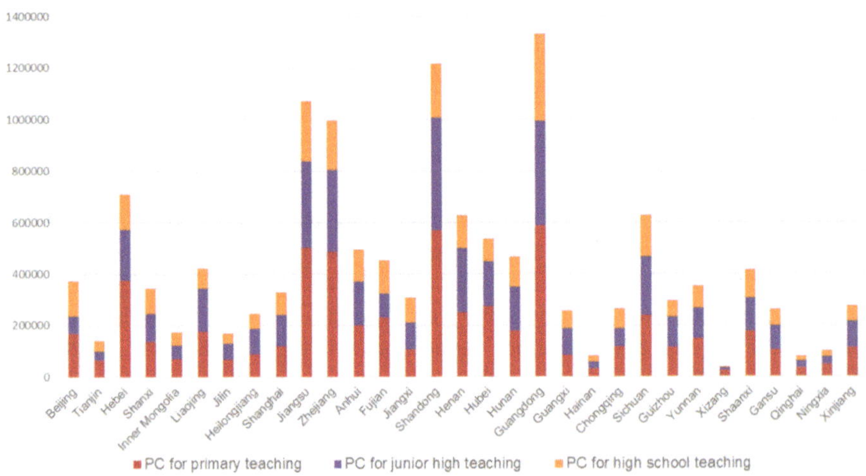

Fig. 4.10 Instructional computer number for each period of basic education at some areas (*unit* Set)

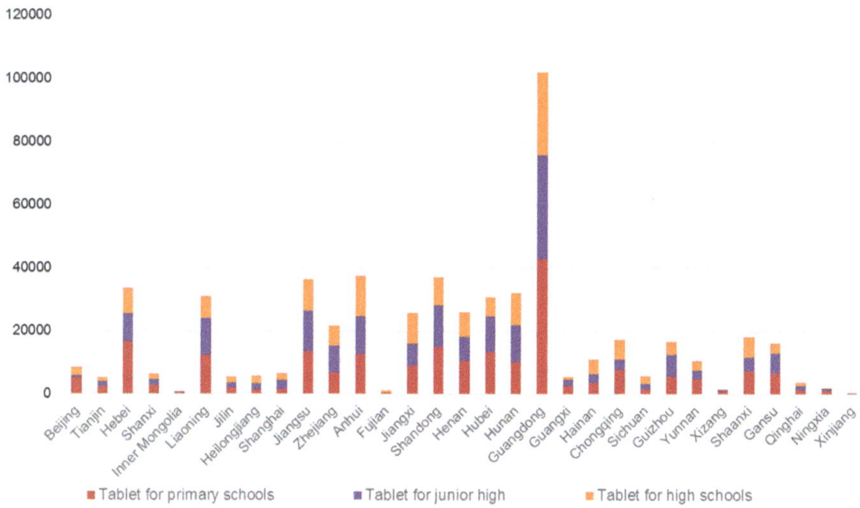

Fig. 4.11 Instructional tablet # for each period of basic education in different provinces

By the end of 2013, the proportion of instructional computers to the number of computers in a school at basic education as well as the proportion of tablet computers to instructional computers is shown in Table 4.1.

Zhejiang Province has the highest proportion of instructional computers to the total number of computers in school during basic education period (87.14%) and the overall proportion nationwide was 79.83%. Guangxi had the lowest proportion of 71.70%. The one with the highest proportion of tablet computers to instructional computers in school was Hainan Province, with 13.50%, and the overall proportion nationwide was 4.19%; Xinjiang was the lowest, which was 0.23%.

During primary school period, the one with the highest proportion of instructional computers to the total number of computers in school was Zhejiang Province, of 89.74% and the overall proportion nationwide was 81.05%. Guangxi was the lowest, which is 63.72%. The one with the highest proportion of tablet computers to instructional computers in school was Hainan Province, with 10.71%, and nationwide was 3.83%; Xinjiang was the lowest, of 0.18%.

During junior high period, the one with the highest proportion of instructional computers to the total number of computers at school was Hubei Province, of 86.98% and the nationwide it was 80.88%. Jilin Province was the lowest, 72.59%. The one with the highest proportion of tablet computers to instructional computers was Hainan Province, with 10.97%, and nationwide was 4.08%. The lowest was Inner Mongolia, which did not equip tablet computers at junior high level.

During high school period, the one with the highest proportion of instructional computers to total number of computers at school was Beijing, which was as high as 84.94%, and nationwide it was 76.29%. Guangdong Province was the lowest, 58.94%. The one with the highest proportion of tablet computers to instructional

Table 4.1 Proportion of instructional computers and tablet computers in basic education at different provinces

Area	Proportion of instructional computer in primary school period (%)	Proportion of tablet computer in primary school period (%)	Proportion of instructional computer in junior high period (%)	Proportion of tablet computer in junior high period (%)	Proportion of instructional computer in high school period (%)	Proportion of tablet computer in high school period (%)	Proportion of instructional computer in basic education period (%)	Proportion of tablet computer in basic education period (%)
Total	**81.05**	**3.83**	**80.88**	**4.08**	**76.29**	**5.03**	**79.83**	**4.19**
Beijing	83.84	3.13	81.83	1.09	**84.94**	1.82	83.88	2.29
Tianjing	83.00	3.85	78.46	4.57	75.16	3.41	79.54	3.89
Hebei	83.89	4.48	82.64	4.51	79.67	5.74	82.70	4.73
Shanxi	83.37	2.22	82.89	1.48	80.60	1.90	82.43	1.89
Inner Mongolia	75.71	0.68	75.81	**0.00**	77.92	**0.00**	76.37	0.28
Liaoning	74.66	7.06	75.34	6.90	71.47	8.83	74.31	7.32
Jilin	71.65	3.09	**72.59**	2.38	73.24	4.69	72.37	3.21
Heilongjiang	78.71	1.56	79.21	2.00	76.22	4.36	78.31	2.39
Shanghai	81.59	1.49	84.65	2.21	82.23	2.51	82.88	2.03
Jiangsu	81.85	2.75	76.12	3.77	77.55	4.27	79.03	3.40
Zhejiang	**89.74**	1.42	86.05	2.70	82.75	3.24	**87.14**	2.18
Anhui	79.83	6.39	79.85	7.07	78.18	10.04	79.41	7.54
Fujian	83.10	0.24	79.85	0.11	77.46	0.45	80.74	0.28
Jiangxi	75.37	8.50	79.11	6.92	80.09	9.61	78.07	8.34
Shandong	80.18	2.64	79.16	2.99	77.95	4.25	79.42	3.05
Henan	82.73	4.19	84.85	3.14	83.00	5.98	83.61	4.14
Hubei	88.52	4.90	**86.98**	6.49	77.16	6.96	85.94	5.75
Hunan	81.67	5.75	82.85	6.85	80.29	8.71	81.74	6.89

(continued)

Table 4.1 (continued)

Area	Proportion of instructional computer in primary school period (%)	Proportion of tablet computer in primary school period (%)	Proportion of instructional computer in junior high period (%)	Proportion of tablet computer in junior high period (%)	Proportion of instructional computer in high school period (%)	Proportion of tablet computer in high school period (%)	Proportion of instructional computer in basic education period (%)	Proportion of tablet computer in basic education period (%)
Guangdong	78.92	7.29	82.11	8.07	**58.94**	7.81	73.46	7.66
Guangxi	**63.72**	3.16	76.17	1.93	76.86	1.12	**71.70**	2.12
Hainan	78.63	**10.71**	82.37	**10.97**	75.61	**20.47**	78.88	**13.50**
Chongqing	77.42	6.67	76.40	4.71	78.65	7.93	77.51	6.54
Sichuan	78.94	0.71	79.04	0.70	80.55	1.54	79.38	0.92
Guizhou	78.87	5.04	84.19	5.80	82.65	6.49	81.70	5.65
Yunnan	75.71	3.27	82.40	2.39	81.55	3.63	79.26	3.06
Tibet	79.26	3.90	74.71	3.89	71.81	5.10	77.32	4.03
Shanxi	83.69	4.22	84.11	3.36	79.11	6.13	82.59	4.44
Gansu	79.75	6.46	81.88	6.68	76.59	5.15	79.68	6.22
Qinghai	81.78	4.36	82.58	4.93	77.46	5.99	81.04	4.91
Ningxia	78.11	2.38	75.65	2.63	72.76	1.07	76.16	2.18
Xinjiang	81.69	**0.18**	81.49	0.24	76.84	0.34	80.46	**0.23**

The biggest and smallest Number in each column, and the line of total were highlighted in bold.

computers was Hainan Province, which reached 20.47%, and nationwide was 5.03%. Inner Mongolia did not equip tablet computers.

2. **Number of computers in urban area, town area and villages at basic education**

Table 4.2 shows the proportions of instructional computers to total number of computers at basic education, as well as the ratios of instructional tablet computers to instructional computers. To view as a whole, during the basic education period, instructional computers at urban-rural areas had the highest ratio. Except junior high schools, the other education period in urban-rural areas also had the highest ratio of instructional tablet computers.

3. **Informatized terminals for teachers and students at basic education level**

By the end of 2014, informatized terminals for teachers and students at basic education level was shown in Figs. 4.2, 4.3, 4.4, 4.5, 4.6, 4.7, 4.8, 4.9, 4.10, 4.11, and 4.12. Among the students, 83.01% of them were equipped with desktop computers, 18.75% were equipped with notebook computers, and 3.333% of them were equipped with tablet computers. Among the teachers, 61.22% of them were equipped with desktop computers, 33.25% were equipped with notebook computers, and 1.43% of them were equipped with tablet computers.

4. **Teacher-computer ratio in basic education**

Teacher-computer ratio at different instructional periods of basic education for different regions is shown in Table 4.3. Generally, during basic education period, school teachers at urban area used computer resources most frequently. During primary school period and the entire basic education period, rural school teachers had the poorest use of computer resources. At junior high and high school periods, town area school teachers had the poorest use of computer resources.

Considering each area's point of view (see Fig. 4.13), at Shanghai, Beijing and Shandong, every one hundred teachers owned over 100 computers, i.e. teacher computer collocation reached or even exceeded 1:1; at Xinjiang, Heilongjiang, Guangxi, etc., every one hundred teachers' computer collocation status was still a little poor.

5. **Student-computer ratio at basic education**

Change of computer numbers per hundred students at basic education level from 1999 to 2014, is shown in Fig. 4.14. At the level of basic education computer deployment level increased annually, and students' average use of computer resources appeared to increase annually.

The number of computers possessed by every one hundred students at different instructional period at urban areas, town areas and villages in 2013 and 2014 is shown in Table 4.4. Whether from an instructional period or area change point of view, the computer deployment level at basic education period was increasing from 2013 to 2014, and use of computer resources by students was also increasing.

Table 4.2 Ratio of instructional computers and instructional tablet computers at each education period of basic education level in urban area, town area and villages

		Instructional PC proportion at primary school (%)	Tablet PC ratio at primary school (%)	Instructional PC ratio at junior high (%)	Tablet PC ratio at junior high (%)	Instructional PC ratio at high school (%)	Tablet PC ratio at high school (%)	Instructional PC ratio at basic education (%)	Tablet PC ratio at basic education (%)
Total		81.05	3.83	80.88	4.08	76.29	5.03	79.83	4.19
Urban Area	Subtotal	81.46	3.88	79.76	4.06	73.67	4.87	78.41	4.24
	Urban-rural area	82.12	3.79	80.83	5.08	53.34	4.43	71.57	4.34
Town Area	Subtotal	81.85	4.01	81.41	4.12	80.11	5.25	81.26	4.35
	Urban-rural area	82.59	4.05	81.88	4.52	80.81	5.96	81.93	4.65
Villages		79.54	3.54	81.79	4.03	80.31	5.12	80.32	3.77

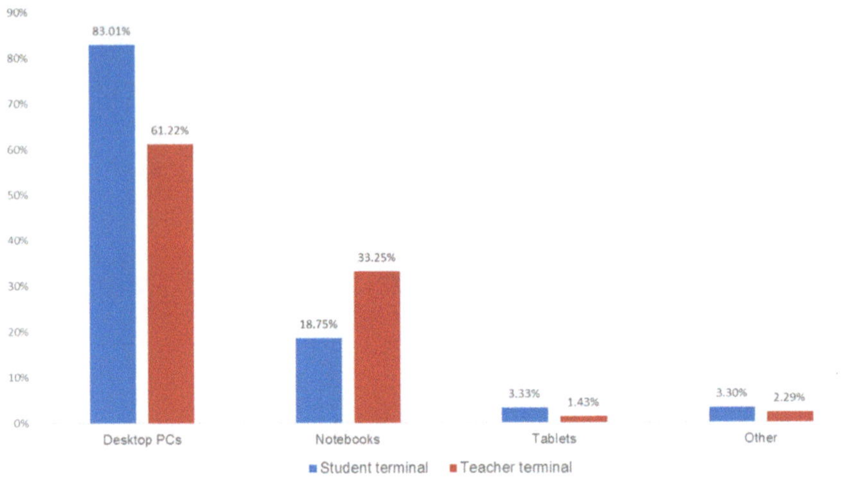

Fig. 4.12 Types and deployment status of informatized terminals for teachers and students at basic education level

Table 4.3 Teacher-computer ratio of each instructional period during basic education period at urban area, town area and villages

	Primary school teacher-computer ratio (Set/Person)	Junior High teacher-computer ratio (Set/person)	High school teacher-computer ratio (Set/person)	Basic education period teacher-computer ratio (Set/person)
Total	1.16	1.26	1.93	1.32
Urban areas	**1.84**	1.53	**2.33**	**1.86**
Urban-rural conglomerates	1.72	**1.62**	2.11	1.77
Town areas	1.11	**1.13**	**1.54**	1.20
Town-rural conglomerates	1.08	1.17	1.56	1.20
Villages	**0.79**	1.20	1.95	**0.92**

The biggest and smallest Number in each column, and the line of total were highlighted in bold.

By the end of 2014, the number of computer owned by every one hundred students at the basic educational level in different areas of the country is shown in Fig. 4.15. At places such as Liaoning, Shanghai, Beijing, Jiangsu, Zhejiang, Shandong, and Ningxia, every one hundred students owned relatively larger number of computers, which almost reached the student-computer ratio requirement stipulated by the national government. In places such as Qinghai, Guangxi, Jiangxi, and Henan, the number of computer owned by every one hundred students was still a little far from the national requirement.

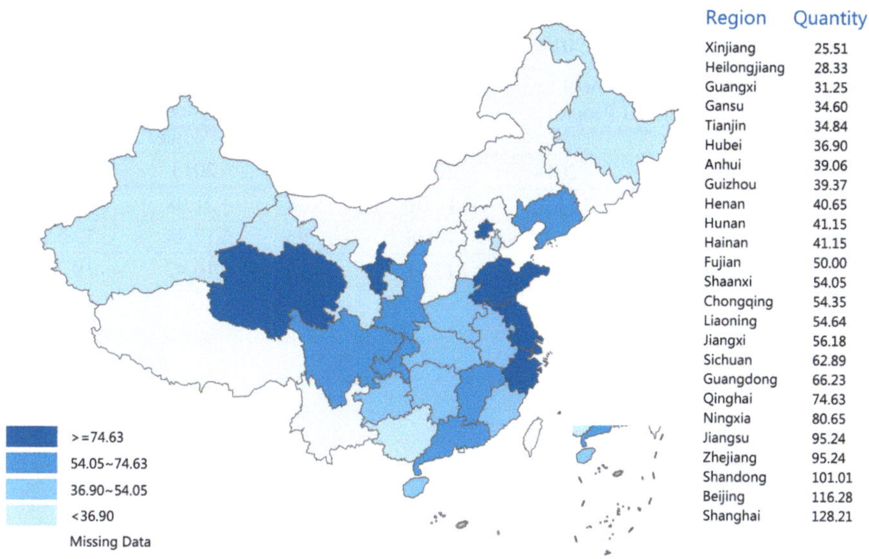

Fig. 4.13 Computer number per 100 teachers have at Basic education for each area (Set/100 people)

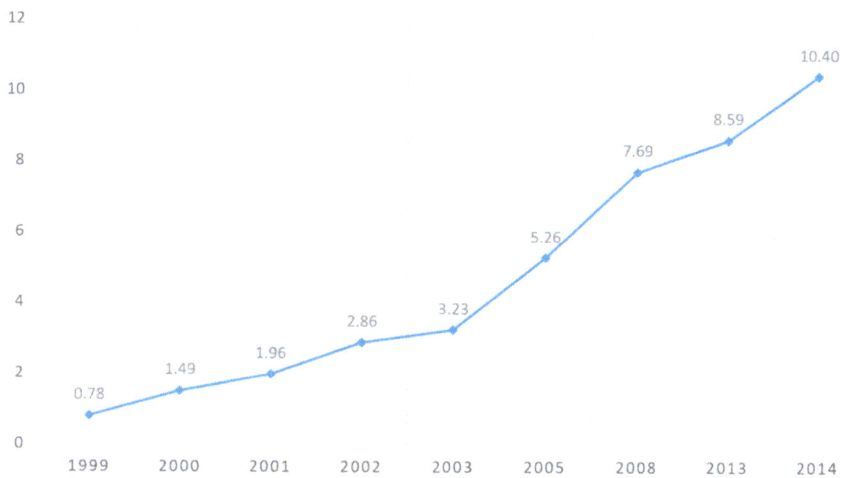

Fig. 4.14 Change of computer number occupancy per 100 students at basic education level (Set/100 People)

Table 4.4 Computer number owned per 100 students at different instructional period at urban areas, town areas, and villages (2013–2014)

	Primary school (Set/100 People)		Junior high school (Set/100 People)		Ordinary high school (Set/100 People)	
	2013	2014	2013	2014	2013	2014
Urban area	10.52	11.22	14.01	15.38	21.98	22.27
Town area	6.98	7.96	10.78	11.95	12.44	13.55
Village	6.36	7.90	13.21	14.73	16.42	19.27

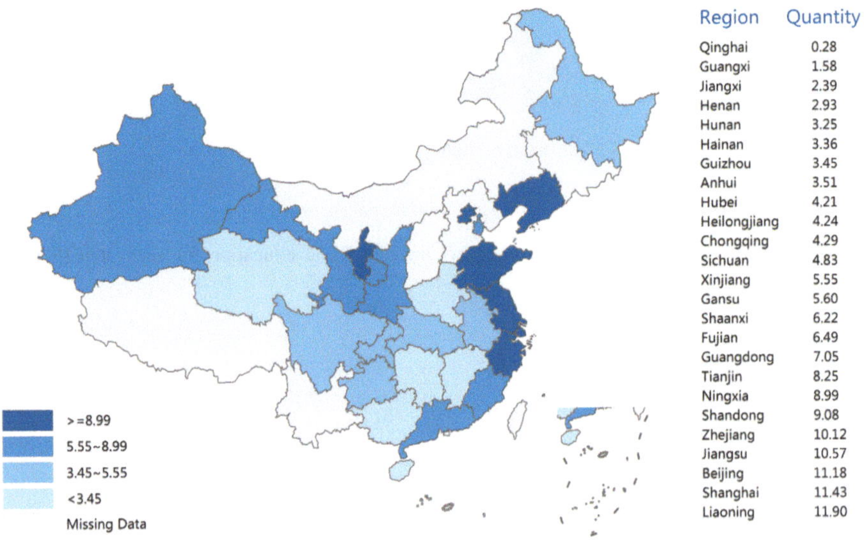

Fig. 4.15 Computer number owned per 100 Students at basic education for each area (Set/100 People)

4.2.2.2 Digital Learning Terminals at Vocational Education

- **Informatized terminals for teachers of vocational education**

The teachers at vocational institutions nationwide were mainly owning desktop computers, notebooks, tablet computers. Figure 4.16 illustrates that 76.56% teachers were equipped with desktop computers, 38% teachers were equipped with notebooks, and 1.32% teachers were equipped with tablet computers; 5.97% of them were equipped with other informatized terminals.

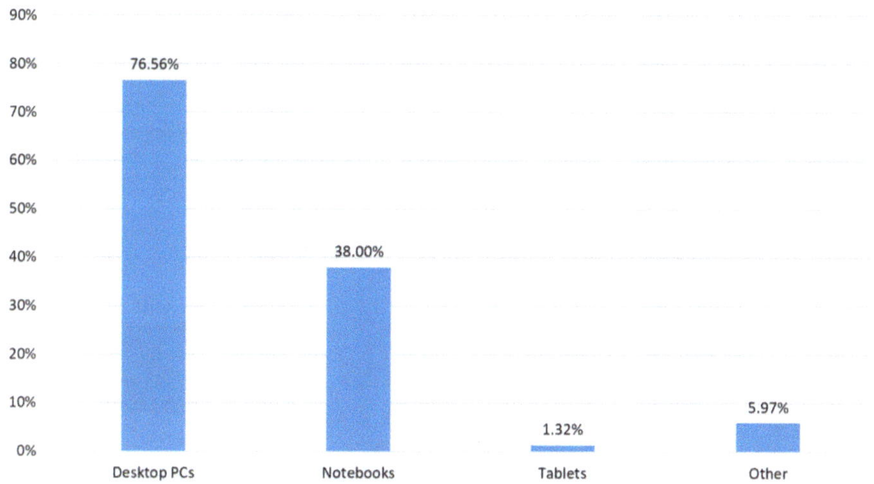

Fig. 4.16 Coverage ratio of informatized terminals of vocational institution teachers nationwide

- **Informatized instructional terminals for teachers owned by vocational institution at different provinces**

The average number of informatized instructional terminals for teachers owned by vocational institutions in different areas of the country is shown in Fig. 4.17. The vocational institutions of Guangxi Province had the highest average number of

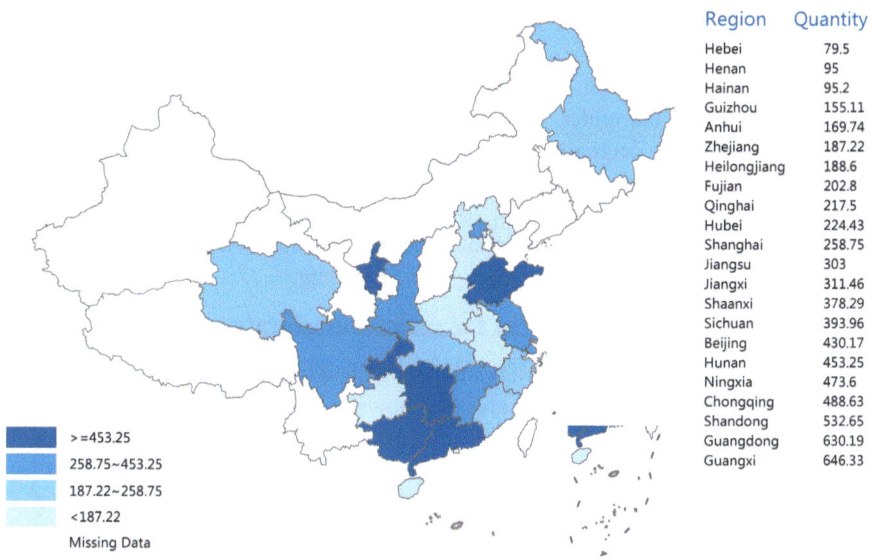

Fig. 4.17 Average teacher terminal number owned by vocational institutions at areas nationwide

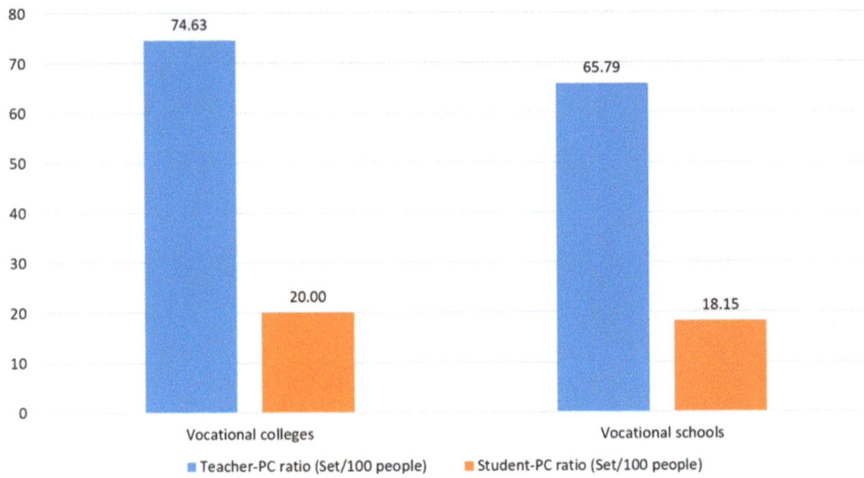

Fig. 4.18 Teacher-PC ratio and Student-PC ratio at vocational colleges and vocational schools

informatized instructional terminals for teachers, and the next were Guangdong, Shandong, Chongqing, Ningxia, Hunan and Beijing.

- **Teacher-PC ratio at vocational institutions**

The teacher-PC ratio and student-PC ratio in vocational colleges and secondary vocational schools by the end of 2013 is shown in Fig. 4.18. The computer resource for teachers and students in vocational colleges was relatively better than secondary vocational schools.

Teacher-PC ratio status of some types of secondary vocational schools (2013, 2014) is shown in Fig. 4.19. In 2014, the teacher computer resource at ordinary secondary technical schools and vocational high schools had some improvement, whereas the teacher computer resource at secondary technical schools for adults and the other institutions decreased.

Number of computers owned by every one hundred teachers at the vocational colleges in 2014 is shown in Fig. 4.20. Shanghai was the best area, for its teacher computer resource status at vocational colleges, and the next were Tibet, Beijing, Shanxi, Zhejiang, Jiangsu, Guangdong and Shandong.

- **Student-PC ratio at each area for vocational education**

By the end of 2014, the number of computers owned by every one hundred students of vocational institutions of some areas is shown in Fig. 4.21. As what was shown for the areas nationwide, Beijing had the best student computer resource deployment in vocational institutions, and the next were Shanghai, Liaoning, Fujian, and Guangdong.

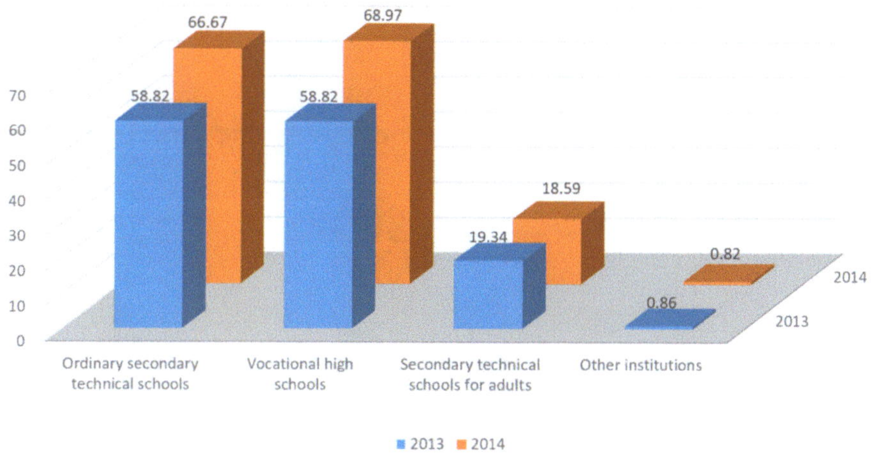

Fig. 4.19 Teacher-PC ratio in some types of vocational schools in 2013 and 2014 (Set/100 People)

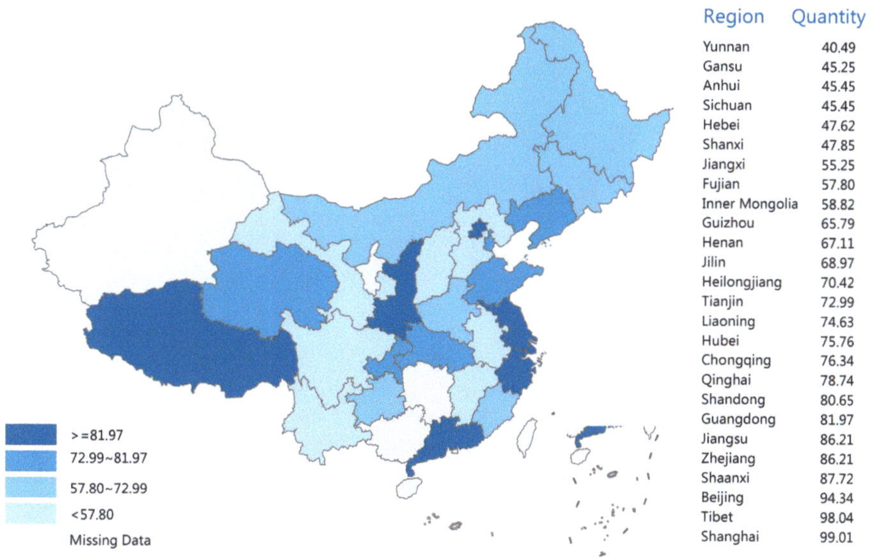

Fig. 4.20 Computer number owned per 100 teachers of vocational education for the areas nationwide (Set/100 People)

4.2.2.3 Digital Learning Terminals in Higher Education

Terminal types equipped for the teachers of colleges and universities were mainly desktop computers, laptop computers, tablet computers etc., and the status is shown in Fig. 4.22. Teachers in institutions of 985 Project were all equipped with desktop and laptop computers, and also there were 28.57% of them equipped with tablet

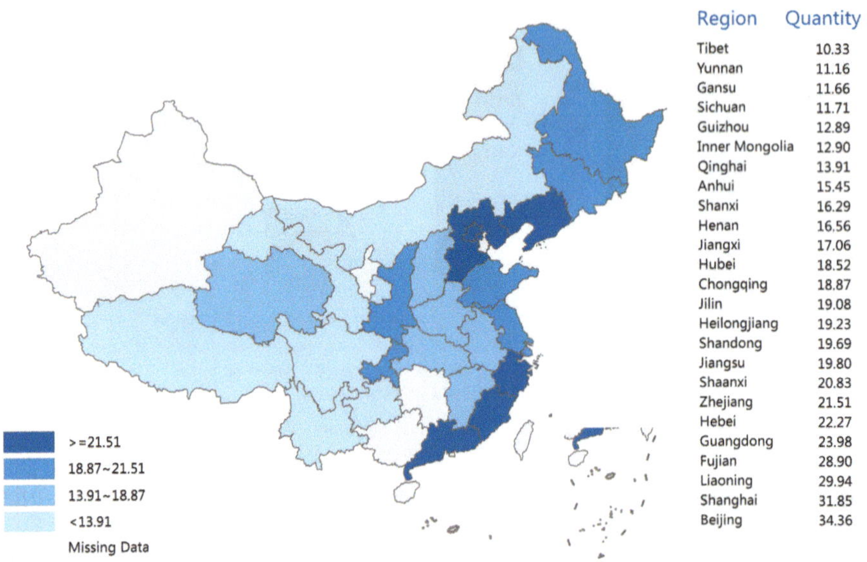

Fig. 4.21 Computers owned per 100 students for vocational institutions of areas nationwide (Set/100 People)

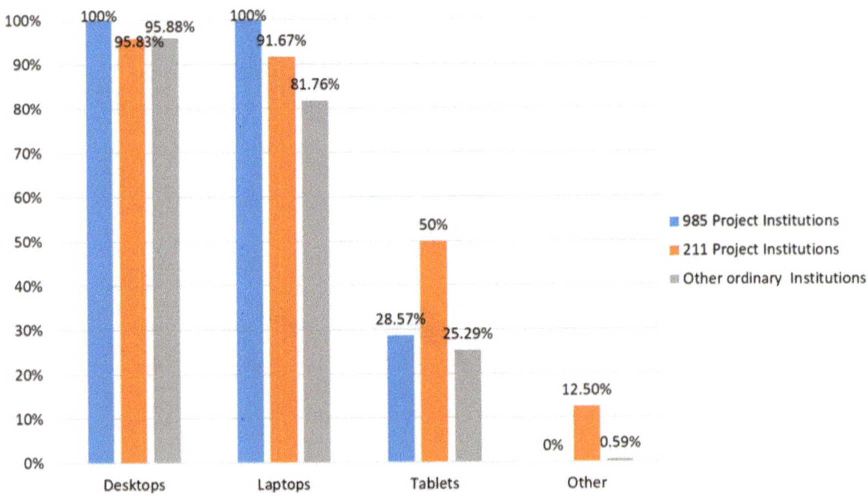

Fig. 4.22 Terminal type and deployment status for higher education teachers

computers. For 211 Project institutions, 95.83% of the teachers were equipped with desktop computers, 91.67% of them were equipped with laptop computers, and there were also 50% equipped with tablet computers. Relatively speaking, teachers in 211 Project institutions had higher allocation rate of tablet computers.

Table 4.5 Total number of computers, instructional computers and tablet computers for higher education institutions nationwide (2013–2014)

Year	Total number of computers		
	Total	Total number of instructional computers	
		Total	Total number of tablet computers
2013	9,720,437	7,400,120	45,816
2014	10,410,006	7,904,544	76,534

Table 4.5 shows the computer number, instructional computer number and instructional tablet computer number for higher education institutions nationwide (including ordinary colleges and universities and adult colleges and universities) in 2013 and 2014.

4.2.3 *Unbalanced Construction of Teaching Support Service System*

Teaching support service system is one of the important components in informatization of education. A good teaching support service system can not only greatly improve work efficiency of teachers and administrators, but also promote innovations in teaching and administration of education. At the current time, informatization of service system's in schools is happening if four ways: (1) It is constructed by the school, and then provided to teachers and students on campus; (2) It is constructed by regional education administrative service uniformly, and provided to teachers and students of primary schools and secondary schools in the region; (3) It is outsourced by schools altogether and entrusted to enterprises for its construction, operation and maintenance; (4) Is packed by the region altogether, and entrusted to enterprises for its construction, operation and maintenance. The text below summarizes a survey on the construction of services supporting teaching according to the demand and characteristics of basic education, vocational education and higher education.

4.2.3.1 Construction Status of Informatization Service System at Basic Education

1. **Campus Smart Card**

Campus Smart Card is one of the symbols of digital campus. It is the foundation of Smart Campus. At primary and secondary schools Campus Smart Card has been implemented mainly as: a dining card, student ID card, library card, time card, and staff card, By the end of 2014, more than 10% of primary and secondary schools implemented Campus Smart Card system. Figure 4.23 shows the general use of Campus Smart Card nationwide at basic education level. We can see that among the

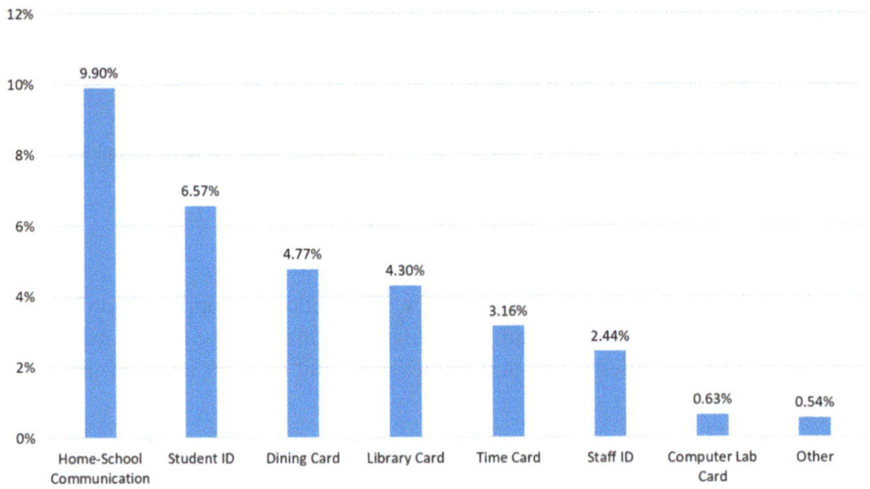

Fig. 4.23 Function implementation status of campus smart card system

schools that utilized Campus Smart Card, it served mainly as student ID, dining card and library card.

2. **Informatization management service system**

The informatization management system often used at basic education level mainly includes: student management information system, library information management system, finance information system, teaching and educational administration information system, facility assets management information system, personnel administration information system, office automation system, file management information system, graduate information system, logistic service information system as well as scientific research information system, etc. Figure 4.24 shows that more than half of the primary and secondary schools were not equipped with an informatized management system. The allocation rate of student management information system was the highest, reached 29.61%, and the next was library information management system.

4.2.3.2 Learning Support Service System at Vocational Education

As shown in Fig. 4.25, among the vocational education institutions nationwide, the ones owning online learning system reached 60% and above, those with a learning resource management platform reached 50% and above, and nearly 15% of them did not have a learning support service system. No more than 50% of the vocational education institutions had a virtual simulated practical training system, learning resource making system, online teaching-research system or online examination system.

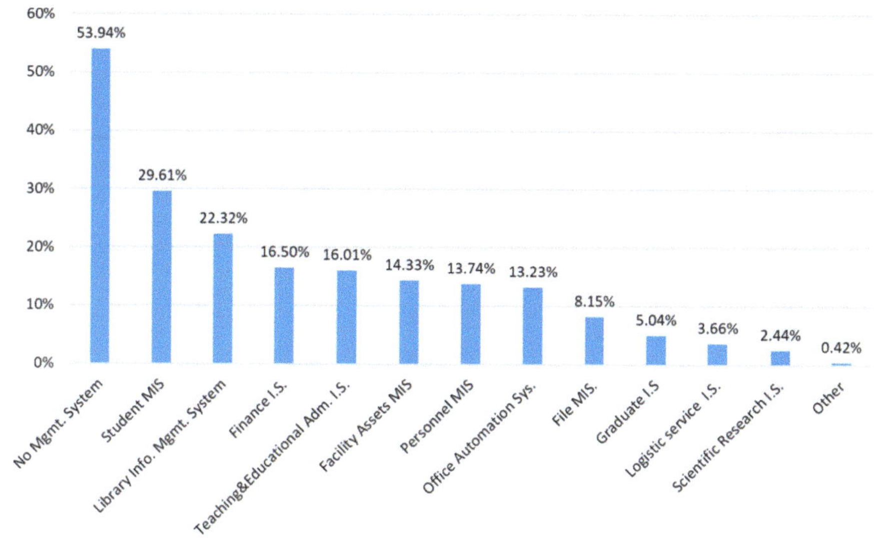

Fig. 4.24 Collocation status of school-level informatization management system

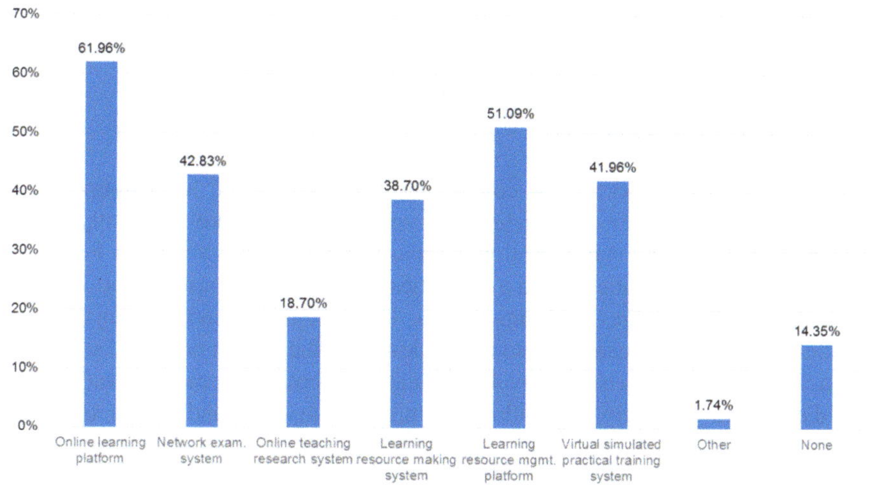

Fig. 4.25 Status of existing learning support service system at vocational institutions

The learning support service systems most often used by students of nationwide vocational education institutions were online learning platform and learning resource management platform. Nearly 50% of students of vocational education institutions used online learning platform, nearly 40% vocational education institutions used learning resource management platform, and there were also many who used virtual practical training system and learning resource making system. The

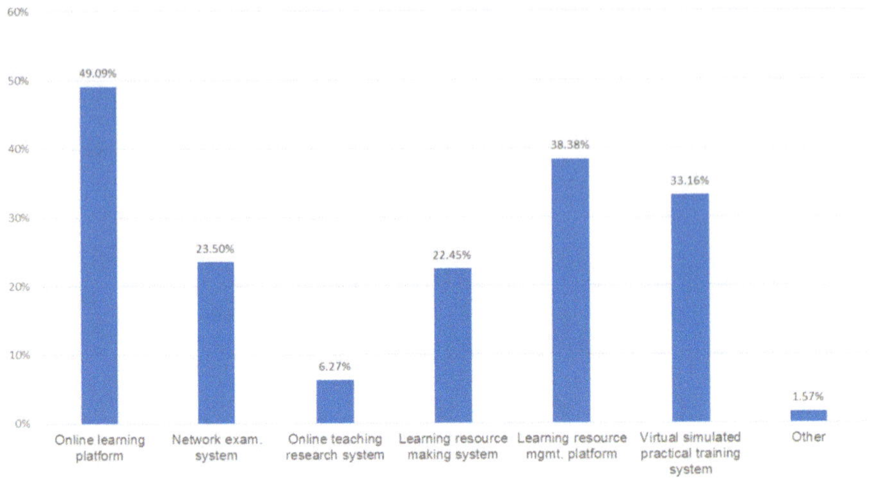

Fig. 4.26 Learning support service system most often used in vocational institutions

online examination system also started to get popular, but the use of online teaching-research system was relatively less. See Fig. 4.26 for detailed information.

4.2.3.3 Construction Status of Learning Support Service System at Higher Education

Colleges and universities of our country have all constructed a learning support service system, and their construction and application level are apparently higher than those at the vocational education and basic education. As shown in Fig. 4.27, in which, above 80% colleges and universities had an online learning platform, and more than 70% colleges and universities had learning resource management. Besides, the system of learning resource making and the system of virtual simulated practical training were also relatively popular, and both reached 40% and above. But there were relatively less online examination systems and online teaching-research systems.

Besides of the learning support service system that supports student learning, the scientific research support service system should also be present at universities and colleges. According to the statistics, by the end of 2014, more than 94% colleges and universities nationwide had uniformly built a scientific research support service system. As shown in Fig. 4.28, more than 50% of colleges and universities had built a literature contribution platform, and the next was that nearly 40% colleges and universities had had a professional tool software platform for scientific research projects.

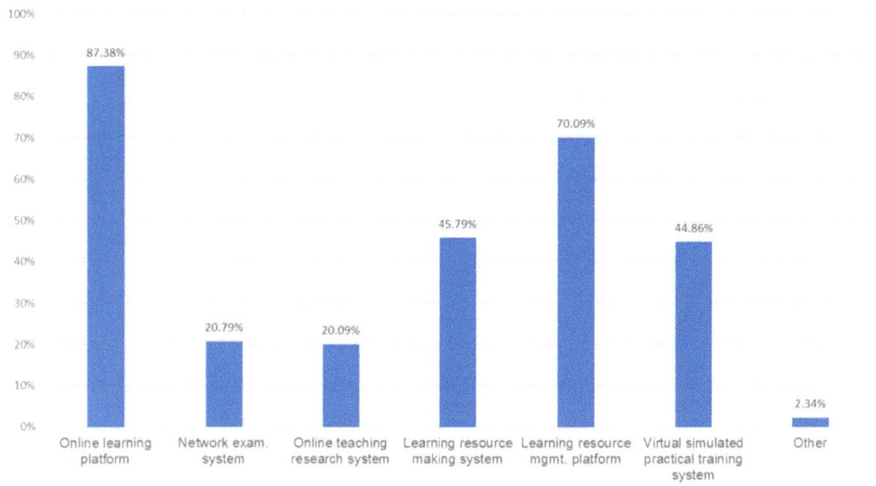

Fig. 4.27 Status of all kinds of teaching informatization systems owned by colleges and universities

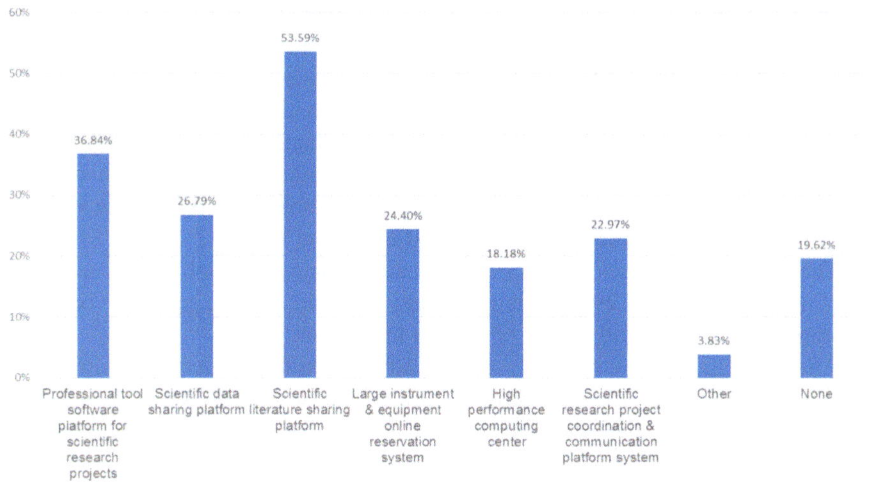

Fig. 4.28 Status of scientific research support service system owned by colleges and universities

4.2.4 Information Safety Has Been Upgraded

Along with the incessant development of education informatization, information safety gets increasing concern. It was clearly pointed out in the *2011–2020 Ten-Year Development Plan of Education Informatization* (MoE 2012), "Enhance

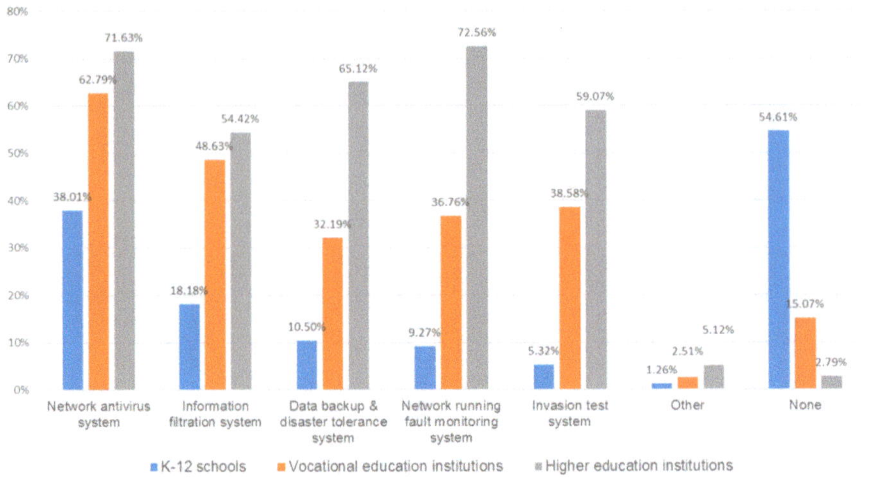

Fig. 4.29 Building status of information security system of all kinds of educational institutions of all levels

safety precaution measures of infrastructure, equipment and information systems; increasingly improve the prevention and emergency response ability to the things such as hostile attack, illegal encroachment etc.; ensure stabilized and reliable operation of the infrastructure, equipment and information systems; Take effective content safety protection measures to prevent dissemination of harmful information; Explore the ways to establish security system and management mechanism for a safe and green informatized environment." This part mainly surveyed the status of school informatized security system at each education level.

The status of information safety systems at all kinds of school at each education level is shown in Fig. 4.29. Differences could be found in K-12 schools, vocational education institutions and higher education institutions.

Among K-12 schools, currently above half of them did not have an information security system, more than one third of the K-12 schools had network anti-virus system, about 10% of the schools had a network running fault detection system, and only about 5% schools had an invasion detection system.

Among the vocational education institutions, by the end of 2014, only around 15% of them had not built an information security system, and most of the institutions had already had an information security system. In which, more than half of the institutions had a network anti-virus system, nearly 50% of institutions had information filtration system, and the construction ratios of network running fault detection system, invasion detection system and data backup and disaster tolerance system were relatively lower.

Among higher education institutions, by the end of 2014, nearly 97% of institutions took multiple protection measures. There were relatively more institutions that had deployed network anti-virus system, data backup and disaster tolerance system, and network running fault detection system; and they all exceeded 65%. The ratios of institutions which had deployed information filtration system or invasion detection system both exceeded 50%. Network information security has been taken more and more seriously by colleges and universities, and they need further improvement in the aspects of data security management and invasion detection.

4.3 Trends in Development of Digital Campus in China

From the data analysis presented in the previous section, we could see that the condition of digital campus has been improved at all kinds of education institutions at all educational levels. In detail, the basic condition for the on-campus network improves notably, teaching terminals are equipped with varied types, the construction of teaching support service system, and network security system receive more and more attention.

The construction of digital campus is the basis for the development of a smart campus. Smart campus refers to a new paradigm of thinking pertaining to a holistic intelligent campus environment which encompasses at least, but not limited to, several themes of campus intelligence, such as holistic e-learning, social networking and communications for work collaboration, green and ICT sustainability with intelligent sensor management systems, protective and preventative health care, smart building management with automated security control and surveillance, and visible campus governance and reporting (Kwok 2015). Construction of smart campus involves mainly key technologies, such as: learning situation identification and environment awareness, campus mobile interconnection, social networking, learning analytics technology, and technology for the organization and sharing of digital resource (Huang et al. 2012a, b).

The gradual maturity of mobile interconnecting technology provides possibility for universal computing, web connection anywhere and anytime, enhance communication and information exchange. Cloud service platform supports online collaboration, file storing, virtualization and flexible visit. These technologies support to use of "seamless connection" on and outside campus, in and outside classroom, at home and at school, which is the basic guarantee to implement "learning at anytime, anywhere, in any way and at any pace (4A)". However, the current status of digital campus still needs to be enhanced to allow learning with the character of DEEE@4A. The mobile interconnecting network and cloud service platform would become the basic configuration in the construction of smart campus. The following three trends will lead the development of digital campus in the next following 10 years in China.

4.3.1 Situation Identifying, Environment Awareness Technology and Social Networking Technologies Will Support the Construction of Smart Campus

Identification of learning situation is the precondition for pushing individualized learning resource to learners, connecting peers for learning groups, as well as providing suggestions to learning activities. It is the key technology in the construction of smart campus. To identify the learning situation one has to: obtain information about location, learning objective, etc., diagnose problems of the learner and predict his/her learning demands, enable the learner to obtain personalized learning resource, find learning companions that can collaborate with each other, and receive effective suggestions regarding effective learning activities. The identification of learning situation involves comprehensive technologies, such as the learner's characteristic analysis, sensor technology and auto deduction.

Technology supporting environmental awareness helps to realize real-time dynamic monitoring, and control all kinds of physical equipment on campus. Awareness technologies such as RFID, two-dimension code, and video monitoring have been used in the areas of campus security, energy saving, scientific research and teaching in schools. For example, the integration of RFID technology with things such as Campus Smart Card, library, instrument, equipment, elevator, and light apparatus can support building entrance management, intelligent checking-into classrooms and meetings, book self-help borrowing and returning, auto inventory checking, expensive equipment anti-theft and positioning, lab open control. It could also make energy-saving control (such as illumination, air-conditioning and ventilation system) more efficient.

According to Boyd and Ellison (2008), social networking refers to the "web-based services that allow individuals to:

(1) construct a public or semi-public profile within a bounded system;
(2) articulate a list of other users with whom they share a connection;
(3) view and traverse their list of connections and those made by others within the system".

The formation and analysis of social networking involves a variety of aspects such as idea, technology, structure and relation. Social networking is significant and global phenomena that has emerged because of the major development of Web applications during the last decade (Eid and Al-Jabri 2016).

Situation identifying technology is the premise for personalized learning resource pushing, learning companion connection as well as learning activity suggestion. Environment awareness technology supports real-time dynamic monitoring, and allows to control all kinds of physical equipment on the campus. Social networking technology helps master activity status of the teachers and students in the virtual network, provides better service for them, and implements necessary regulatory measures. Therefore, intelligent technology allowing identifying the situation environment awareness, and social networking will become critical in the construction of Smart Campus.

4.3.2 *Learning Analytics Based on Big Educational Data Will Become Indispensable*

Learning analytics is the measurement, collection, analysis and reporting of data about learners and their contexts, for the purpose of understanding and optimizing learning and the environments in which it occurs (Brown 2011). Learning analytics technology can be an effective supporting tool for teachers' teaching decision-making, and teaching optimization. It can provide effective data support for students' self-oriented learning, learning crisis pre-warning and self-evaluation, and it can also provide data reference for education researchers' individualized learning design and increase of research benefits.

The core of learning analytics is to build a student model according to the obtained information, such as the learners' personal information, learners' situation information, learners' participation level about learning, learners' social networking. Text mining technology can dig from the learning resource library and learner information all kinds of information that learners pay attention to, such as text topics, viewpoint inclination of text author to something, and professional level of the author to some topic etc. This information can help learners retrieve learning resource based on this information, it can also help learners know about the general situation and trend of the learning resource library.

Connecting the different functional systems of smart campus, collecting big educational data at all educational levels, performing scientific learning analysis based on the data, are the basis to improve teachers' teaching method, increase students' learning initiative and improve the quality of learning. Learning analytics will also help schools and teachers to effectively adjust courses and assessment system, and provide guarantee for the realization of students' smart learning. Therefore, learning analytics based on big educational data will become an indispensable function in the building of Smart Campus.

4.3.3 *Creating Open Learning Environments Is Inevitable*

Organization and sharing of learning resource have always been one of the focal points in the construction of a digital campus. According to Bandura's social learning theory, the observation of other people's behavior is the important origin of acquiring knowledge and skill, and real examples can play exemplary roles to the observers (Bandura 2001). The generative knowledge that learners generate during the learning process could be used as the resources for the other learners. For the convenience of retrieving and sharing, traditional learning resource often uses static glossaries to describe metadata, but this descriptive method cannot satisfy the versatile and extremely personalized learning demand in a ubiquitous learning environment.

The key idea of education is the cultivation of innovative capability for all students. Education should become "open" in time and space to all learners to improve their learning experience. Smart campus supports the exploitation of resource environment, and lets student break through the limitations of textbooks. Smart campus also breaks the wall of time and space to make learning expand from in-class to outside the class. It supports the exploitation of space environment, and makes effective learning able to occur in real situation and virtual situation.

4.4 Ten Best Practices in Constructing of Digital Campus

A group of educational institutions have already emerged in China in with their digital campus implementing intelligent technologies. This section chose ten educational institutions at different educational levels which carry out smart learning by adopting of intelligent technologies.

4.4.1 Exploration of Innovative Multiple-Campus Management Model Based on Smart Campus: Second Primary School of Yudong in Banan District of Chongqing

The Second Primary School of Yudong is located at the Banan District of Chongqing. In 2002, the school was recognized as the "Demonstrative Primary School of Chongqing", a trial unit among the "First Batch of Trial Units for Education Informatization of the Ministry of Education", one of the "A Hundred Demonstrative Schools for E-campus Building Project Nationwide", and a school in the "First Batch of E-campus Demonstrative Schools in the City of Chongqing". The key character of the school is the management of multiple campuses by using ICT.

The Second Primary School of Yudong has three independent campuses. These campuses have already been built up a variety of systems, such campus network, TV, sound-control, recording and broadcasting, and video monitoring. The school started to explore the management mode of smart campus with multiple campuses in 2011, and since then has fully implemented digitized campus management. The school purchased a set of copyrighted "Tongda OA Office System" (Office Anywhere).

A multi-campus integrated digitized campus management system was formed, which uniformly manages the interfaces, unifies databases, and integrates gateways. It also implements informatized management of multiple campuses in the areas such as digitized office work, teaching, educational administration, moral education, scientific research, personnel management, logistics, training and learning, resource, etc. Using the network as a base, the school connected its three campuses with their multimedia equipment such as high-definition video cameras, wireless speakers, and overhead projectors. It can hold visualized meetings hosting staff at three campuses at the same time, with speech given on one campus and watched by the other two, All three campuses can directly communicate and dialogue through the cyber video conferencing systems.

4.4.2 Integration of Lesson Preparation, Instruction, and Teaching-Research Based on Cloud Platform: Sandaojie Primary School in Wuhan of Hubei Province

The broad use in teaching and learning of new technical means such as cloud computing technology and e-textbooks promotes exploration of new forms of teaching, makes formal learning and informal learning more effective.

The Sandaojie Primary School in Wuhan of Hubei Province currently has 22 instructional classes, with 56 teachers and 888 t students. The school created a set of lessons prepared by using cloud and lesson-giving platform shared by teachers and students together. The resource library is constructed here by the way of "school-enterprise cooperation, application-driven and resource-sharing", and realized classroom teaching and learning practice based on the education cloud for teachers and students, as shown in Fig. 4.30.

4.4.3 Implementation of Flipped Classroom in All-Subject Based on Online Teaching System: First Changle High School of Shandong Province

The First Changle High School of Shandong Province was founded in 1938. It is located at the southeast of the downtown of Changle County of the City of Weifang. It is a key high school of Shandong Province. First Changle High built an online teaching system comprised of reciprocal digitized learning platform, campus image server, 1000M campus network and personal tablet computers used by students. The digitized learning platform includes recording of student learning,

Fig. 4.30 Solution of Sandaojie Primary of Wuhan of Hubei province for lesson preparation, lesson giving, and teaching-research integration based on cloud

feedback, analysis, and evaluation system. The online teaching and learning system, is supported by the advantages of flipped classroom, the method of collective lesson preparation, guided learning of lesson plans, and cooperative group learning, First Changle High school created the teaching model of "two sections, four steps and ten links", and implemented flipped learning into all subjects within the whole school, as shown in Fig. 4.31.

4.4.4 The Integrated Service Platform of Educational Administration and Teaching: Affiliated High School of Beijing Institute of Technology

The Affiliated High School of Beijing Institute of Technology was founded in 1950. In 1980, it was recognized as the key high school in Haidian District. In 2004, it became a demonstrative regular high school of Beijing.

The school has successively established an integrated educational administration and teaching service platform that includes intelligent class scheduling system, online course selection system, online test-paper scoring system, grade analytic system, and teaching evaluation system. In September 2013, the school established

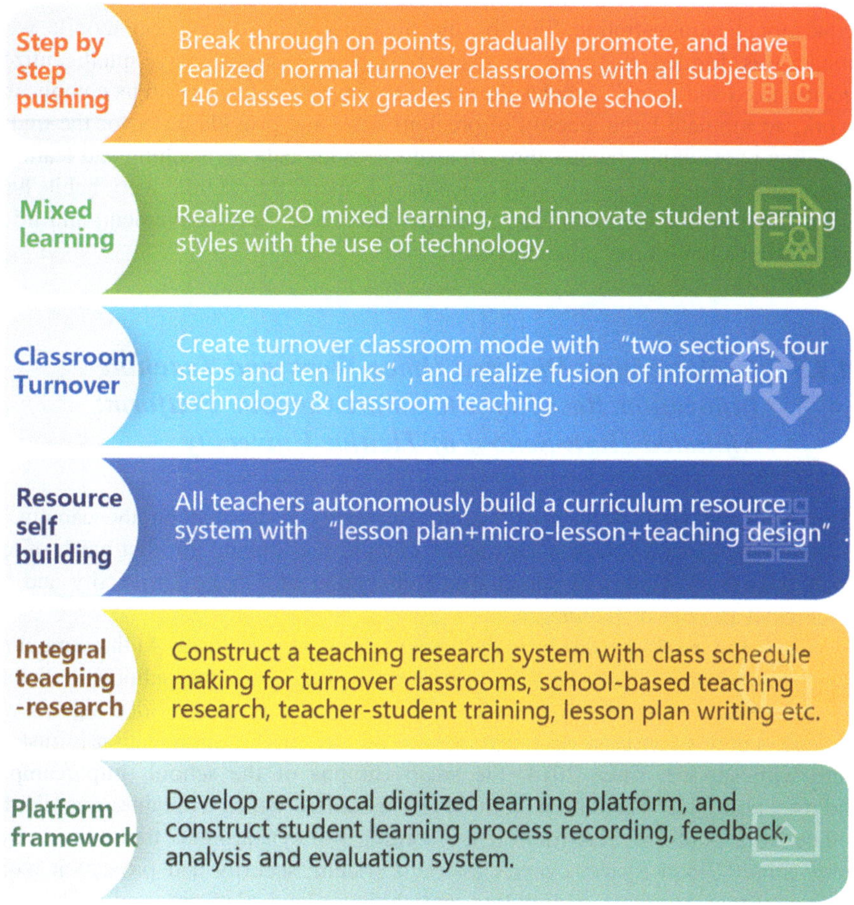

Fig. 4.31 Flipped classroom teaching based on online learning system of first changle high of shandong

micro-course platform building workgroup, set up especially as a teacher micro-course studio, and started the design and development work of micro-course platform and micro-courses. Currently it has set up over twenty courses, with total browsing amount of 5655 times, and the highest browsing amount of single micro-lesson video of 857 times.

Students choose learning content on their own according to their learning preference and the instruction of their teachers with the micro-course community. They exploit new ways of learning from the teachers, classmates and other people through the network, and change from passive learning to independent learning.

At the same time, the school established office applications such as scheduling plan, mail receiving and sending, on-campus instant messaging, personal cloud disk, information bulletin, personnel management, student status management, and educational administration. Finally, the digitized, networking, and mobile work mode was formed that can be characterized as "omnibearing, intelligentized, economized and green". It helps the school to simplify and speed up its complicated work, accumulate large amount of precious data, and provide basis for the understanding of teaching changes through analysis of the data on teaching and learning. Through digitized construction of office management, the original work habits have been changed, the communication became faster and more convenient, and information transmission became smoother.

4.4.5 Implementing Digitized Office Work and Teaching Innovation Based on the Smart Campus Platform: Affiliated High School of Tianjin University

The Affiliated High School of Tianjin University is located on the campus of Tianjin University in Nankai District of Tianjin. It is among the first batch of key high schools of Tianjin under dual administration of Tianjin University and the Municipal Bureau of Education.

In 2013, wireless coverage on the whole school was provided. At the same time, all kinds of online resources in school were integrated, the school's e-School Administration was initially created, and the total coverage of video signals was almost completed. Based on integration of the original informatization infrastructure with service, since 2014, the smart campus of the school (http://campus.tjdxfszx.com) has been built with seven modules, which contains: coordinated office work platform, classroom real-recording and synchronous transmission system, Smart Cloud Space, cyber classroom system, security and protection monitoring system, resource co-building and sharing cloud platform, and e-file folder system used to record students' growth.

Through the Smart Campus administration platform, the school can not only fully implement digitized office work and improve work efficiency, but also provide school administration with data for decision-making based on scientific analysis. The classroom real-recording, synchronous transmission system, resource co-building and sharing cloud platform provide effective technical support for the collection of high quality teaching resources through real-time recording. The smart cloud space and cyber classroom system provide conditions for obtaining of high quality learning resource at anytime and anywhere. With the learning platform and the e-portfolio of students, big data about student' learning can be formed. Analysis of the big data can help teachers, parents and students to clearly know about the current learning situation and existing problems, as well as offer relevant instruction and guidance.

4.4.6 Construction of Smart Campus of "One Platform and Two Systems" Helps with Teaching and Administrative Innovation: Yantian High School of Shenzhen of Guangdong Province

Yantian High School of Shenzhen is the only regular and public key high school in Yantian District of Shenzhen. The school listed the building of a smart campus as the major project in its five-year development plan, and carefully designed its development scheme of "one platform and two systems". This includes a smart learning and teaching platform, smart management decision-making system and smart service system.

The school currently has set up a smart campus C/S platform that includes grade management system, student status management system, course selection system. The school has also built cloud service platform, teaching resource library system, teacher evaluation system, Smart Card system (including checking-in, consumption, entrance guard, water and electricity control etc.), intelligent broadcasting system, security monitoring system, online grade analytic system, student growth recording system and apartment management system. The intelligentialized education and teaching platform collecting the big data generated during teaching activities, and through the learning analytic technology, can increase the effectiveness of teaching decision-making. The informatized management decision-making system cares about students' growth and teachers' professional development, increases school's information analysis ability, provides basis for scientific decision-making of management, and improves administrators' work efficiency and management skill. The smart service system provides convenient campus life service and home-school interaction service.

4.4.7 Implementing Innovations in Administration of Teaching and Talent Cultivation with the Support of Smart Campus Ecology: Changzhou College of Information Technology of Jiangsu

Changzhou College of Information Technology is located at the city of Changzhou of Jiangsu Province. It is the first college of information technology in Jiangsu Province. It belongs to Jiangsu Economic and Information Technology Commission, and is one of the one hundred demonstrative vocational colleges for the whole country.

Focusing on users and closely addressing teachers' and students' needs, Changzhou College of Information Technology has built up a Smart Campus ecology with emphases on the web portal of the college, five platforms ("Sunshine College Administration" Integrated Management and Service, Online Learning,

Energy Saving, Safety, and Harmonious Campus), and two centers (Online Teacher Development, and Student Service). With the support of Smart Campus ecology, Changzhou College of Information Technology successfully implemented application of MOOC (Massive Open Online Course), real-time update of status data of talent cultivation, real-time update of the information such as registered permanent residence for the students, archives, and employment, as well as the building and application of micro-community based on WeChat platform.

4.4.8 Smart Education Administration and Web Learning Platform Based on Cloud Service and Virtualization Technology: Shandong Polytechnic College

Shandong Polytechnic College is a regular vocational college with over sixty years of school operation history, which was approved by the People's Government of Shandong Province, and has been registered at China's Ministry of Education. In 2014, it was listed among the first batch of Trial Units of Education Informatization of Shandong Province.

The college invested over 10 million RMB to build a campus network with 10,000M, double core, and 1000 million RMB to desktop and realized network environment with full wired and wireless coverage. It also built a series of management application systems such as office automation, Smart Card and teaching administration, and realized informatization and digitization of the campus.

Cooperating with enterprises utilizing technologies such as virtualization and cloud, the college has built data center and service center based on Cloud Service. Besides, it will collaborate with relevant colleges and enterprises to build a professional teaching resource library, connect professional teaching resource at secondary and higher education level, and to realize sharing of teaching resource with the colleges and enterprises. Finally, it will create an integrated construction of One Campus (digitized campus), One Library (Shared Digitized Teaching Resource Library), and One Platform (Smart Education Management and Online Learning Platform).

4.4.9 Improving Teaching, Scientific Research and Management Based on the Internet of Things: Nanjing University of Posts and Telecommunications

Nanjing University of Posts & Telecommunications is located in the city of Nanjing of Jiangsu Province. It is a university jointly built by China's Ministry of Industry and Information Technology and People's Government of Jiangsu Province. Nanjing University of Posts and Telecommunications is one of the universities that work on the research of The Internet of Things and building of Smart Campus the earliest in China. In 2009, the university took the lead to plan to build a Smart Campus based on The Internet of Things among the domestic higher education institutions. And its demonstration project (first phase) of Smart Campus passed expert evaluation in August 2013.

Based on the newly-emerged technologies, such as Cloud Computing, virtualization and The Internet of Things, the university has built an intelligent campus environment geared to the needs of teaching, scientific research, administration and campus life. With the integration of environment awareness, network fusion and open intelligent application, the smart campus provides comprehensive information service to teachers and students, which make teachers and students be able to quickly and accurately access information and service on campus about people, property, things, as well as the processes of production, learning, and research. Besides, the university reformed the management improvement and work flow in the smart campus, and took it as an important content in the university's system innovation and administration innovation. With intelligent comprehensive data analysis, the smart campus provides the most basic data support for all kinds of decision-making of the university, and realizes scientific decision-making. Through close connection among all application systems in the smart campus, it provides on-campus resource-sharing, information sharing, information transmission and information service, thus improves teaching quality, scientific level and management level.

4.4.10 Providing One-Stop Service for Teachers and Students Based on Mobile Smart Campus: Guizhou Institute of Technology

Guizhou Institute of Technology is located in the city of Guiyang of Guizhou Province. It is a regular provincial technical institute founded with the approval of the Ministry of Education. Since 2013, at the beginning of the establishment of the institute, it took the building of digitized campus as a highlight project in its informatization construction.

After two years of construction, in July 2015, Guizhou Institute of Technology build and opened the first Smart Mobile Campus platform within the province. The platform has two kinds of applications on mobile end and PC end respectively.

The PC terminal is a typical digital campus comprehensive service portal, which provides One-Stop service for teachers and students, with the integration of learning management system, educational administration system, library system, campus Smart Card system, finance system, personnel system, mailbox system, and assets system. Its mobile end is the first mobile Smart Campus platform in Guizhou Province which is fully opened. It provides all kinds of application services such as information aggregation, data inquiry, and business handling, and integrates APPs of all kinds of functions. It transmits information at any time about campus news, notifications and announcements, campus activities, course information, life consumption, book borrowing and reading, and mobile OA (Office Automation). Besides, the mobile end will also continuously release applications, such as empty classroom inquiry, academic lectures, grade inquiry, and provide standard developer service. This will allow the institute developing relevant applications with its own developer team, and provide service to all the teachers and students of the institute. The mobile Smart Campus platform will gradually assist the institute's teachers and students and realize study and work at any place and any time.

References

Bandura, A. (2001). Social cognitive theory: An agentic perspective. *Annual Review of Psychology, 52*(1), 1–26.

Boyd, D., & Ellison, N. B. (2008). Social network sites: definition, history, and scholarship. *Journal of Computer-Mediated Communication, 13*(1), 210–230. (Bukowitz)

Brown, M. (2011). Learning Analytics: the coming third wave. *EDUCAUSE,* (April), 1–4. Retrieved from http://scholar.google.com/scholar?hl=en&btnG=Search&q=intitle:Learning+Analytics:+the+coming+third+wave#0

Eid, M. I. M., & Al-Jabri, I. M. (2016). Social networking, knowledge sharing, and student learning: The case of university students. *Computers & Education, 99*, 14–27. doi:http://dx.doi.org/10.1016/j.compedu.2016.04.007

Huang, R. (2009). The strategic emphasis and construction focus of digital campus. *Beijing Education (Basic Education), 07,* 6–7.

Huang, R., Zhang, J., Hu, Y., & Yang, J. (2012a). Smart campus: The developing trends of digital campus. *Open Education Research, 18*(4), 12–17.

Huang, R., Zhang, J., Hu, Y., & Yang, J. (2012b). The concept and characters of smart classroom. *Open Education Research, 18*(2), 12–17.

Kwok, L. (2015). A vision for the development of i-campus. *Smart Learning Environments, 2*(1), 2. doi:10.1186/s40561-015-0009-8

MoE of P.R.C (2012), *Ten-year development plan for ICT in education in China (2011–2020).* Retrieved from http://www.moe.edu.cn/ewebeditor/uploadfile/2012/03/29/20120329140800968.doc83c4d329.html

MoE of P.R.C. (2014). *Educational Statistics 2014. Website of MOE, China.* Retrieved from http://www.moe.edu.cn/s78/A03/moe_560/jytjsj_2014/

Wu, D. (2015), *Development report about Chinese education informatization.* Beijing: Beijing Normal University Press.

Chapter 5
Smart Learning in Classroom Environment

Abstract Smart classroom is a typical environment for smart learning, and it is the high end form of digital classroom. This chapter is presenting the process of creating smart classrooms in China. Firstly, the trends of ICT in classroom is analyzed. Secondly, the status of multiple types of classroom (multimedia, computer, interactive white board and tablet computer classrooms) is analyzed. Thirdly, ways to build smart classroom are discussed from classroom equipment, spatial arrangement, teaching application, and cases.

Keywords Smart classroom · Learning experience · Construction of classroom · The trend of smart classroom

5.1 Introduction

The immediate task for education is to embrace the future, and empower our children to learn with the cultural tools they have already been given (Soloway 1991). In China, a batch of education ICT-based projects has been launched successively since 2000, and great development has been achieved in the area of creating educational system based on ICT infrastructure. Since 2010, more attention was paid to building classroom learning environment based on ICT. Governmental policies have been proposing that the classroom environment supporting technology will be the important element of ICT-based education. Starting from 2000, ICT has been placed on the education reform agenda both as an object of education, and as an important pedagogical tool for innovative teaching methodology (Peeraer and Van Petegem 2015). As argued by Williams (2005), extrapolations from emerging trends can have a value in promoting understanding of the present situation, as well as in identifying needs for the future. ICT is one of the most visible symbols of globalization and educational innovation (Power 2007). Implementation of ICT is often presented as 'inevitable' (James and Hopkinson 2009) and increasingly ubiquitous. The integration of ICT with education is moving beyond getting

D. Liu et al., *Smart Learning in Smart Cities*, Lecture Notes in Educational Technology, DOI 10.1007/978-981-10-4343-7_5

personal computers into the hands of learners and towards mobile technology, virtual world, and cloud computing amongst others (Hong and Songan 2011).

Each generation, has a distinctive 'generational consciousness', which is dependent on the pace of social change (Erstad 2010). The implications of such generational divides are further popularized in Tapscott's book 'Growing up digital. The rise of the Net generation' (1998), and with a recent follow up called 'Grown up digital' (2008), and a similar book by Palfrey and Gasser termed 'Born digital' (2008). Digital generation of learners is good at utilizing technology and pay attention to teamwork and practical operation, and prefers new learning methods, such as autonomous learning, exploration and cooperation, pursuing for innovation (Huang et al. 2014). The features of smart learning, such as smart perception, strong interaction and high immersion, earn appeal to digital learner's learning needs (Huang et al. 2012). However, the traditional classroom environment is not beneficial to the development of smart learning, and the learners have relatively low satisfaction from learning experience in current classroom environment. The performance level required from the smart classroom is far more complex than the one required in traditional classroom teaching (Ringstaff et al. 1991).

In the information age, with the sufficient development of sensor technology, network technology, media technology and artificial intelligence technology, the classroom environment becomes the "net type of classroom which can optimize teaching content presentation, be convenient for learning research obtaining, promote classroom interaction development, with context awareness and environment management function". Smart classroom is a typical materialization of smart learning environment, and is the high end form of network classroom, where the "intelligence" involves five dimensions: Showing, Manageable, Accessible, Real-time interactive and Testing, abbreviated as "SMART" conceptual model, shown as Fig. 5.1 (Huang et al. 2012).

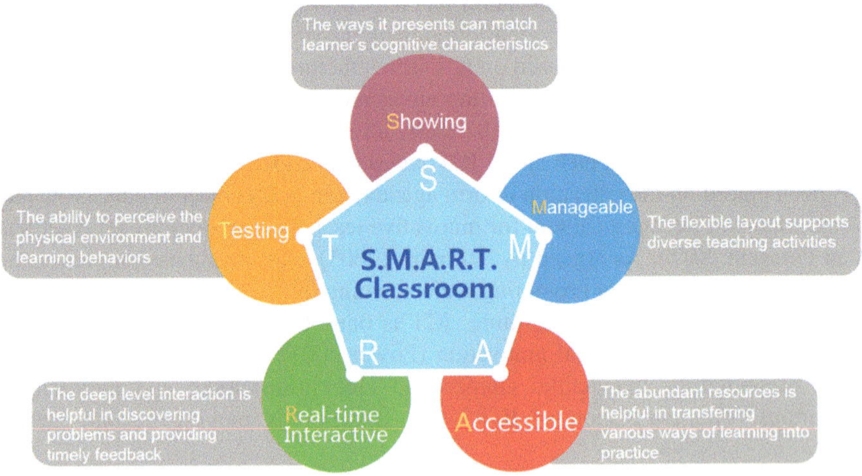

Fig. 5.1 The concept of "SMART" classroom

With the development of smart classroom, some trends emerged in the building digital classrooms in China, which will be illustrated in the next section.

5.2 The Trends of ICT in Classroom

The teaching environment will change with the development of smart classroom. Since the highly development of Internet, teaching content has change from the traditional one-way teaching into the bidirectional interaction, teaching environment will change from the vivid field teaching into the virtual classroom, and teaching contents will also change from single to multiple applications. Traditional classrooms often involve several drawbacks due to limitations of formal learning (Novosadova et al. 2007). With the development of information technology, as well as mobile devices, the transformation of the education and teaching has been changed, especially with the appearance of flipped classroom, the concept of smart classroom has been already introduced in more and more countries. Smart classroom can promote the construction of traditional classroom. In the coming years, we infer that there will be five trends under the influence on the development of smart classroom.

5.3 Applying Innovative Artificial Intelligence in Education: Learning Companion and Smart Classrooms

A growing number of educational studies apply sensors to improve student learning in real classroom settings. However, how can sensors be integrated into classrooms to help learners improve their learning effectiveness? One way to solve this problem with AI is learning companion with sensor technologies. Timms envisions smart classroom that make use of sensors to support learning and illustrates how they might be used in new ways if AIED (Artificial Intelligence in Education) applications are embedded into them (Timms 2016). Learning companions provide both task-related feedback and motivational support for students with the goal of improving learning, such as education robot (Lubold et al. 2016). As robots become increasingly pervasive, filling every aspect of life, at home, at work, at school, they can offer continued and individualized support in cases where it is not always possible to have a constant human companion. With this growing number of applications involving social human-robot interactions, there is a growing need for adaptive and personalized learning. Through cloud computing, network technology and response systems and other technology, smart classroom can support the development of adaptive and personalized learning.

5.4 Transforming Teaching Mode

The future perspectives of smart learning environments are reviewed and shared, through examples of emerging innovations such as the flipped classroom, game based learning, gesture base learning, along with pedagogical shifts, such as life-long learning portfolio maintenance, team teaching, and separation of learning and competency assessment. (Kinshuk et al. 2016). Therefore, in order to improve the students' learning autonomy, the current teaching mode need to be changed, such as the task driven teaching mode, which was increasingly recognized and promoted by colleges and universities. Duanyuan et al. give an example of task—driven teaching automation specialty, they show that task—driven teaching can integrate more dispersed knowledge in a teaching task that requires students make integrated use of key knowledge points to complete the task, thus effectively mobilize the participation of students' consciousness (2016). Task-driven teaching mode is very suitable for the smart classroom. Project-based teaching mode also can be integrated into smart classroom. The history of PBL can be traced back to the progressive tradition advocated by John Dewey. Dewey insisted on the idea of "learning by doing". He argued that the classroom should be a kind of society and the students should be encouraged to become the center in the learning process. Project is an extended task which usually integrates skills through a number of activities (Hedge 1993). Project-based learning (PBL) is yet another way technology can be utilized to facilitate STEM learning in ways that align with real-world practices. With the change of smart classroom, the traditional teaching mode also need to be changed.

5.5 Preparing Teachers and Students for Smart Classroom

In a smart classroom, the role of the teacher has changed from being a performer to a designer and director of learning situations. This new situation is also a challenge for lecturers at teacher training programs. Now they have to design learning situations promoting practical skills of the students in the pre-service training to make them became teachers working according to the new curriculum and technologies (Mattsson and Mutvei 2015). European Commission (2006) defines competence as the combination of knowledge, skills and attitudes that a person must have to successfully develop and participate in the knowledge society. Teachers also need to consider employing various strategies to enhance positive relationships with students in order to facilitate more effective learning experience and vary strategies for enhancing satisfactory relationship with students depending upon different learning environment (Song et al. 2016). Both students and teachers need a change in their teaching ability and learning ability, thus learn how to make good use of technologies is a key point. Future more, teacher training for smart classroom is also in a hurry, they need to know how to keep up with the new technologies and teaching mode.

5.6 Combining STEM with Smart Classroom

Technology use in the classroom has great potential to transform student learning. This is particularly true for science, technology, engineering, and mathematics (STEM) workplace applications of technology, such as robotics, computer modeling and simulations, digital animation, multimedia production, biotechnology, and geospatial technologies (Parker et al. 2015). These applications offer opportunities to align classroom activities with real-world STEM practices, while engaging students in authentic investigations and design-based pursuits and promoting student-centered learning and interest in STEM careers (MaKinster et al. 2014; Miller et al. 2011). Technology applications can provide opportunities for tackling these high-quality STEM practices through interactive learning opportunities that mirror professionals' use of digital tools (Moeller and Reitzes 2011). Integration of STEM workplace applications of technology into K–12 smart classrooms presents significant challenges for education.

5.7 Integrating Maker into Smart Classroom

The development of technologies such as 3D printing, open source electronics, and accessible programming environments has given rise to a despecialization of aspects of prototyping and production so that broader populations can participate in 'do-it-yourself' (DIY) activities once reserved for technology experts (Chu et al. 2015). This phenomenon has spurred interest in the use of Making in educational settings to teach such STEM (Science, Technology, Engineering and Mathematics) curriculum content as science, engineering, electronics, robotics, and mathematics, and to encourage children to pursue STEM careers later in life. Making oriented learning needs not be constrained to the assessment of specific skills and knowledge sets. Maker address problems by seeking 'do-it-yourself' solutions, a belief that they either have or can acquire the means to construct a solution, and possess a creative curiosity to seek solutions that the Maker can construct on her own (Chu et al. 2015). Martinez and Stager describe Making as the "act of creation with new and familiar materials" where one takes control of their life and their learning (Martin and Dixon 2013). In order to combine making oriented leaning and smart classroom together, the school need a set of equipment, a specific method, a bounded place, or even a community of practitioners. It is a whole culture that celebrates certain key values of personal production and problem solving.

The development of classroom environment is closely related to the policies in our country. For example, the "Shanghai Medium and Long Term Educational Reform and Development Planning Outline (2011–2020)" mentions that in five years Shanghai plans to develop and construct the instructive, interactive and systematic digital teaching system beneficial for self-regulated learning and. By the end of March of 2015, the implementation of "connect classes[1] with quality digital learning resources" had an obvious effect: 73% of the schools have owned multimedia classrooms, and 43% of the schools created multimedia classroom. Other provinces and regions have the same situation. The next section will introduce the overall status of classroom environments in China with abundant data analysis.

5.8 The Status of Classroom in Chinese Schools

To understand the status of classroom environment, we will consider two factors. Firstly, we will analyze the issue of hardware equipment, as the perfect and complete equipment is the basis for a digital classroom. Secondly, we will analyze how schools develop cloud service system, as availability of network resources is beneficial for students.

The types of classroom environment include multimedia classroom, computer room, interactive white board classroom and tablet computer classroom. Therefore, the status of digital classroom in this chapter is described from four aspects: multimedia classroom, computer room, interactive white board classroom and tablet computer classroom.

The data about multimedia classroom, computer room and tablet computer classroom are from the 2014 educational statistics released by the Development Department of the Ministry of Education of the People's Republic of China.

The data about interactive white board classroom are from the "Chinese Electronic Whiteboard Market Analysis Report", released in recent years by China Market Research Group. The sales data of whiteboard from the report are utilized to conduct analysis on the implementation of interactive white board classrooms in our country.

[1]The concept "Connect Classes" means equipping the classes with moderate ICT-based equipment and network facilities, and equipping schools with corresponding ICT-based teaching resources. Connect Classes is implemented as one of the important policies mentioned in "National Plan for ICT in Education (2011–2020)".

5.9 Implementation of Multimedia Classroom in Chinese Schools

5.9.1 Multimedia Classroom in Primary School

The coverage rate of multimedia classrooms of primary schools in various regions throughout the country is shown in Fig. 5.2. The coverage rate of multimedia classrooms in primary schools of Zhejiang, Beijing and Shanghai is relatively high, more than 90%, and the coverage rate of Gansu, Xizang, Guangxi and Henan is lower than that of the average level of the whole country.

5.9.2 Multimedia Classroom in Junior High School

The coverage rate of network multimedia classrooms of junior high schools in various regions throughout the country is shown in Fig. 5.3. The coverage rate of network multimedia classrooms in junior high schools of Shandong, Shanghai, Beijing and Zhejiang is relatively high, higher than 80%, and the coverage rate of Xizang, Sichuan and Hunan is relatively low.

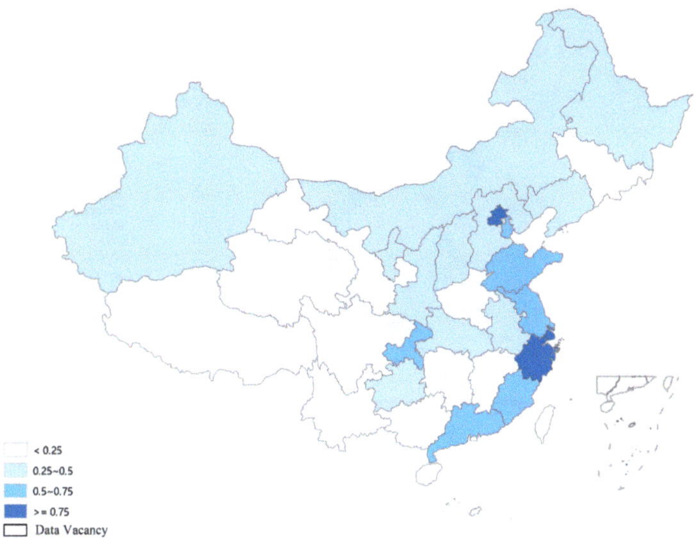

Province	Rate
Gansu	15.04%
Tibet	17.20%
Guangxi	17.68%
Henan	18.85%
Jilin	18.86%
Yunnan	20.65%
Hunan	21.17%
Hainan	22.11%
Jiangxi	22.51%
Sichuan	22.90%
Qinghai	23.36%
Shaanxi	27.31%
Hebei	27.84%
Heilongjiang	28.33%
Guizhou	36.11%
Liaoning	37.15%
Inner Mongolia	40.65%
Ningxia	40.91%
Anhui	41.52%
Shanxi	44.38%
Hubei	46.47%
Xinjiang	48.95%
Guangdong	51.67%
Fujian	54.61%
Tianjin	58.28%
Chongqing	59.96%
Shandong	65.52%
Jiangsu	70.39%
Shanghai	90.80%
Beijing	92.33%
Zhejiang	96.90%

Fig. 5.2 Coverage rate of network multimedia classroom of primary schools of various regions throughout the country

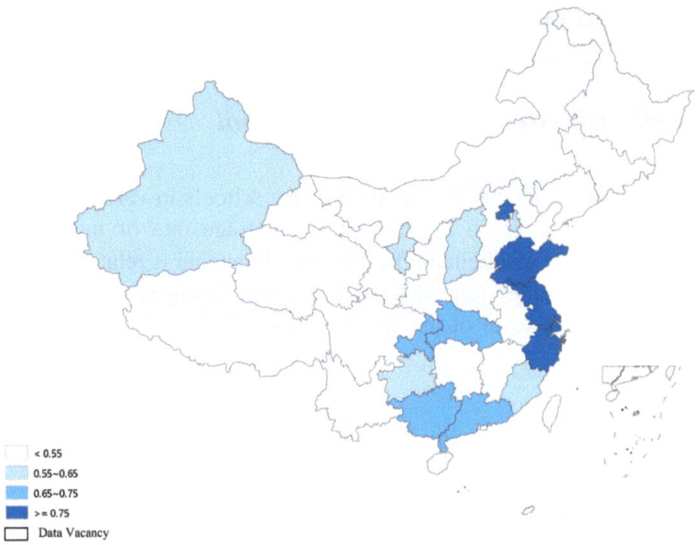

Province	Rate
Tibet	21.54%
Sichuan	32.78%
Hunan	35.41%
Qinghai	35.75%
Jilin	37.30%
Gansu	38.58%
Shaanxi	38.84%
Jiangxi	41.20%
Heilongjiang	41.76%
Hainan	41.94%
Henan	42.46%
Liaoning	45.72%
Hebei	49.70%
Yunnan	49.91%
Inner Mongolia	49.91%
Anhui	54.39%
Guizhou	55.65%
Tianjin	56.20%
Shanxi	56.58%
Xinjiang	57.75%
Ningxia	58.12%
Fujian	63.48%
Guangxi	65.45%
Huibei	67.41%
Guangdong	69.13%
Chongqing	73.86%
Jiangsu	78.51%
Shandong	80.68%
Shanghai	85.68%
Beijing	89.44%
Zhejiang	94.61%

Fig. 5.3 Coverage rate of multimedia classrooms in junior high schools of various regions throughout the country

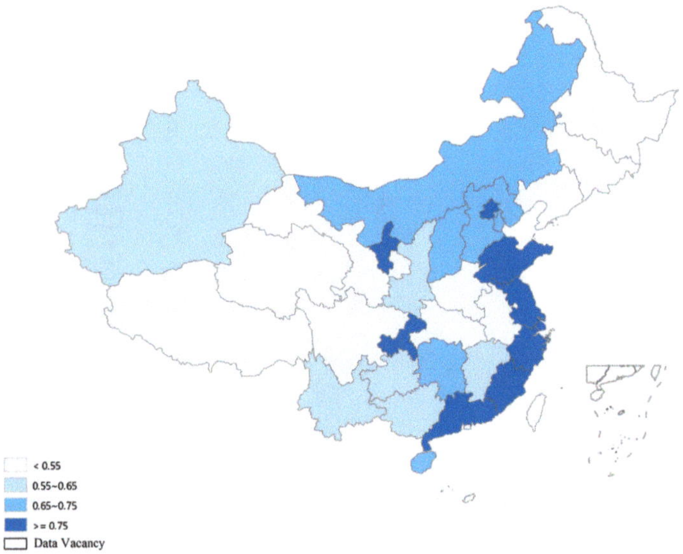

Province	Rate
Tibet	33.31%
Qinghai	45.54%
Anhui	46.70%
Sichuan	49.24%
Henan	50.60%
Liaoning	51.56%
Heilongjiang	51.89%
Gansu	52.13%
Jilin	52.72%
Hubei	54.26%
Shaanxi	55.82%
Jiangxi	56.52%
Xinjiang	57.52%
Guizhou	58.24%
Yunnan	58.25%
Guangxi	60.62%
Heibei	66.15%
Hainan	66.54%
Shanxi	66.62%
Inner Mongolia	70.88%
Hunan	71.84%
Tianjin	74.38%
Guangdong	76.41%
Fujian	77.68%
Shandong	79.14%
Chongqing	79.53%
Ningxia	83.73%
Jiangsu	84.62%
Shanghai	87.88%
Beijing	92.32%
Zhejiang	94.24%

Fig. 5.4 Coverage rate of network multimedia classrooms of regular senior secondary school of various regions throughout the country

5.9.3 Multimedia Classroom in Senior High School

The coverage rate of network multimedia classrooms of senior high schools in various regions throughout the country is shown as Fig. 5.4. The coverage rate of network multimedia classrooms of ordinary senior high schools of Zhejiang, Beijing, Shanghai, Jiangsu and Ningxia is relatively high, higher than 80%, and the coverage rate of Xizang, Qinghai and Anhui is relatively low, lower than average level of the whole country.

5.10 Implementation of Computer Classroom in Chinese Schools

The number of computer rooms reflects the basic ability of schools to utilize ITC-based means to develop teaching activities. According to the 2014 educational statistical data[2] released by Development Department of the Ministry of Education of the People's Republic of China, there are great differences of the computer room area occupied by each 10 thousand students between various regions and learning stages.

5.10.1 Computer Classroom in Primary School

The Fig. 5.5 illustrates the space used by computer rooms per 10 thousand students of primary schools. In Ningxia and Shaanxi, the space is bigger (with the area of 2013 m^2 of Ningxia and 2004 m^2 of Shaanxi), while in Guangxi and Jiangxi the space is relatively smaller.

5.10.2 Computer Classroom in Junior High School

Figure 5.6 illustrates the space used by computer rooms per 10 thousand students of junior high schools in various regions. In Jiangsu and Shanghai the space is bigger (4218 m^2 of Jiangsu and 3455 m^2 of Shanghai), while in Guangxi and Guizhou the space is relatively lower.

---

[2]Data sources: http://www.moe.edu.cn/s78/A03/moe_560/jytjsj_2014/.

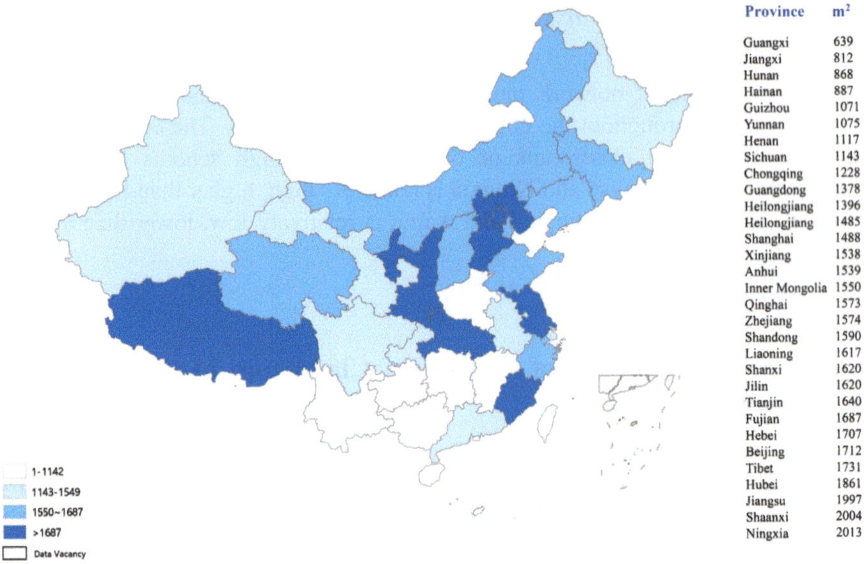

Fig. 5.5 Computer room occupation area of 10 thousand students of primary schools in various regions (m^2/10 thousand people)

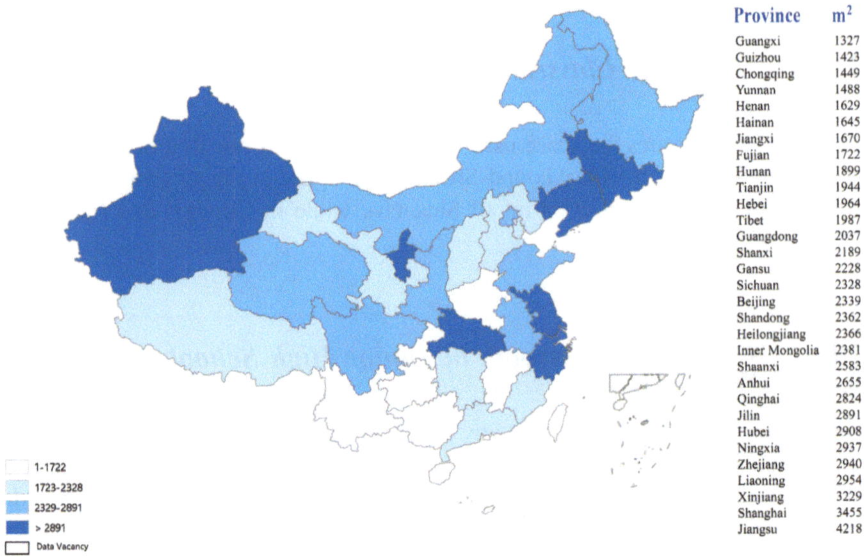

Fig. 5.6 Computer room occupation area of 10 thousand students of junior high schools in various regions (m^2/10 thousand people)

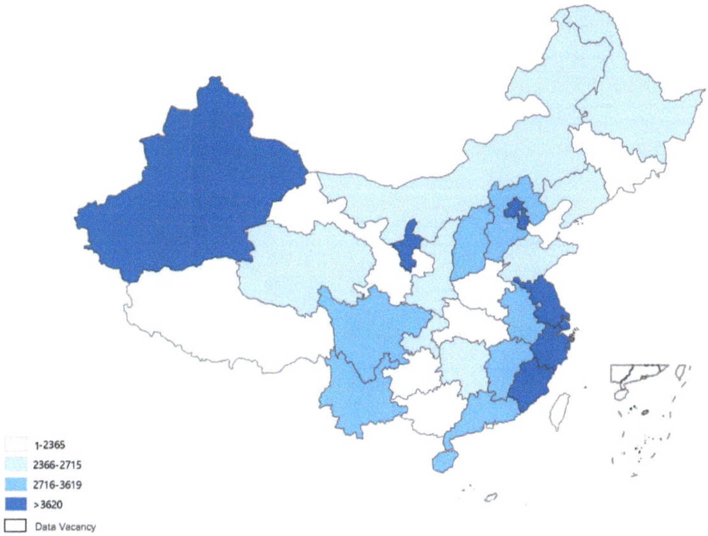

Province	m^2
Tibet	1200
Guizhou	1671
Henan	1873
Jilin	2030
Guangxi	2034
Hubei	2224
Gansu	2354
Heilongjiang	2365
Shandong	2445
Shaanxi	2465
Liaoning	2529
Inner Mongolia	2544
Chongqing	2587
Qinghai	2648
Hunan	2685
Sichuan	2716
Yunnan	2743
Anhui	2786
Jiangxi	2886
Hebei	2945
Guangdong	2947
Hainan	3203
Zhejiang	3620
Ningxia	3709
Xinjiang	3848
Fujian	4129
Tianjin	4246
Jiangsu	5897
Shanghai	6679
Beijing	7318

Fig. 5.7 Computer room occupation area of 10 thousand students of senior high schools in various regions (m^2/10 thousand people)

5.10.3 Computer Classroom in Senior High School

Figure 5.7 illustrates the space used by computer rooms per 10 thousand students of senior high schools in various regions. In Beijing and Shanghai the space is relatively higher (7318 m^2 of Beijing and 6679 m^2 of Shanghai), while in Guangxi and Guizhou the space is relatively lower.

5.11 Implementation of Interactive White Board Classroom in Chinese Schools

The construction of interactive white board classrooms is the embodiment of the implementation of ITC-based education into construction of digital classroom environment. It reflects the basic ability of schools to utilize ITC-based means in teaching activities. In consideration that 95% of the white boards sales in our country is used for education, we analyzed the sales data of electric white boards, presented in "Analysis Report to Electric White Board Market Development in China" released in 2012–2014.

We found out that the total holding quantity of electric white boards in various regions of our country is not high, and there is obvious difference between the holding quantity of interactive white boards in various regions.

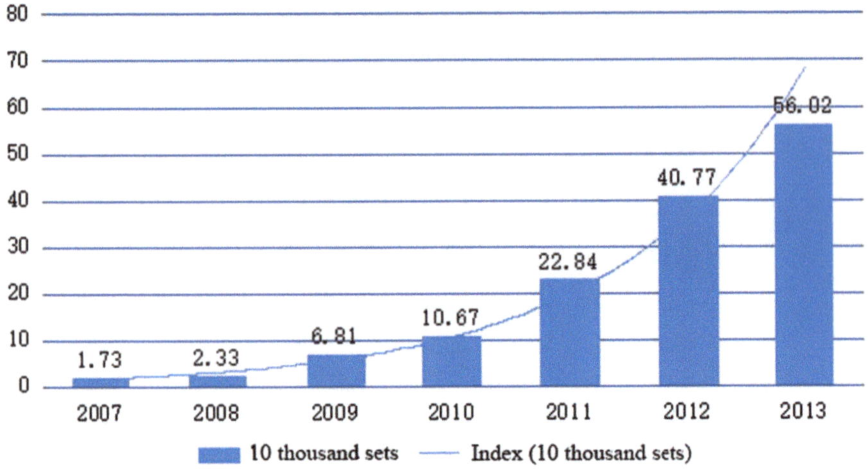

Fig. 5.8 Nationwide sales of electric white boards in the period of 2007–2013

5.11.1 Interactive White Board Classrooms Throughout the Country

The amount of interactive white board classrooms in our country has gradually increased, with qualitative change from 17.3 thousand sets in 2007 to 560.2 thousand sets in 2013, as shown in Fig. 5.8.

According to the statistics from the year 2013, based on the quantity of electric whiteboards used for education, and the quantity of multimedia classrooms in 2013, the proportion of the multimedia classrooms and interactive white boards is relatively low (about 1/3), as shown in Fig. 5.9.

Fig. 5.9 Proportion of multimedia classrooms with interactive white boards in 2013

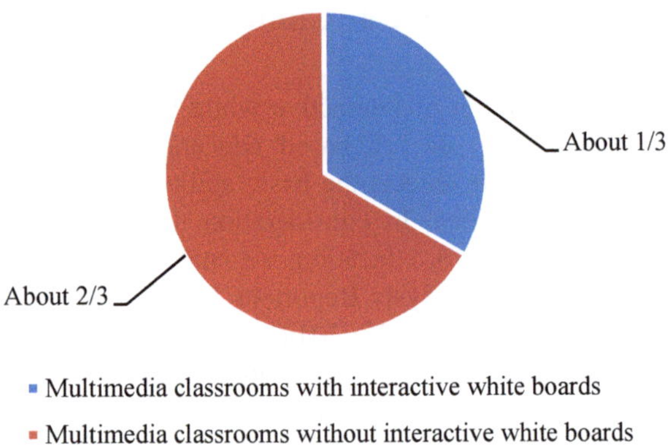

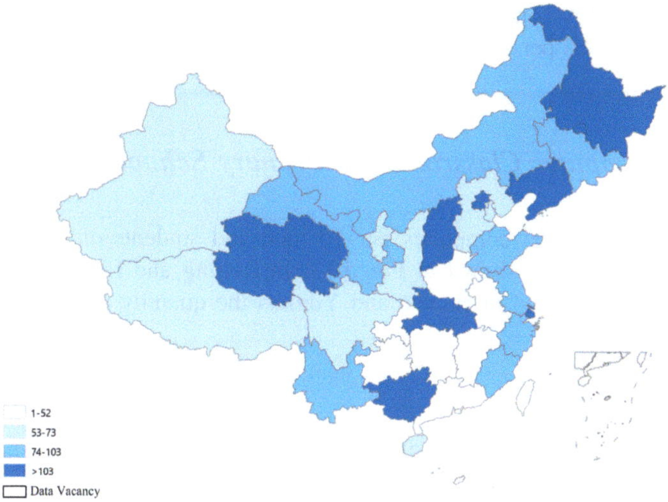

Province	Quantity
Jiangxi	29
Chongqing	30
Guizhou	31
Anhui	45
Henan	47
Hunan	49
Guangdong	52
Tibet	52
Ningxia	53
Hebei	58
Shaanxi	66
Hainan	66
Xinjiang	67
Tianjin	67
Sichuan	70
Jiangsu	70
Shandong	72
Jilin	82
Gansu	89
Fujian	89
Inner Mongolia	96
Zhejiang	96
Yunnan	99
Guangxi	106
Heilongjiang	106
Qinghai	106
Shanxi	118
Hubei	141
Liaoning	151
Shanghai	214
Beijing	294

Fig. 5.10 Quantity of interactive white board classrooms owned by 10 Thousand students in various regions

5.11.2 Interactive White Board Classrooms in Different Regions of China

The use of interactive white board classrooms in various regions of the country is shown in Fig. 5.10. The quantity of interactive white board classrooms per 10 thousand students in Beijing and Shanghai is relatively higher (Beijing is the highest), and the quantity of interactive white board classrooms per 10 thousand students in Jiangxi and Chongqing is relatively lower.

5.12 Implementation of Tablet Computer Classrooms in Chinese Schools

Equipping classrooms with tablet is the embodiment of the intelligent classroom environment. Considering that the smart classroom construction in our country is still in exploratory and developmental stage, the quantity of tablet computer classrooms is relatively small. We analyzed the data about the quantity of tablet computer published in 2014 education statistic data[3] released by Development Department of the Ministry of Education of the People's Republic of China.

[3]Data sources: http://www.moe.edu.cn/s78/A03/moe_560/jytjsj_2014/.

We found out the quantity of tablet computers in various regions of various leaning stages in our country is not sufficient, and there is obvious difference of the quantity of tablet computers per 10 thousand students between various regions.

5.12.1 Tablet Computer Classrooms in Primary School

The quantity of teaching tablet computers per 10 thousand students of primary schools in various provinces is shown in Fig. 5.11. In Beijing and Liaoning the quantity is relatively bigger, and in Hainan and Yunnan the quantity is relatively smaller.

5.12.2 Tablet Computer Classrooms in Junior High School

The quantity of teaching tablet computers per 10 thousand students in junior high schools in various regions is shown as Fig. 5.12. In Beijing the quantity is relatively bigger, and in Inner Mongolia and Yunnan the quantity is relatively smaller.

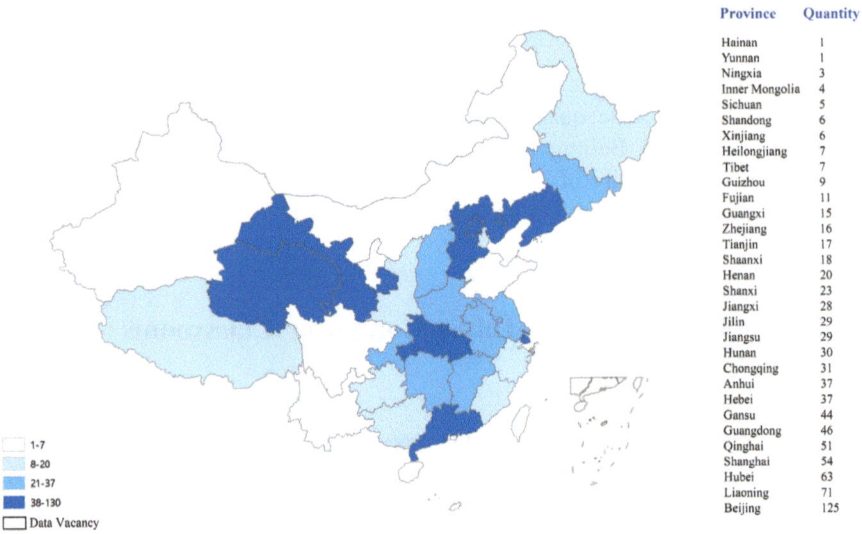

Fig. 5.11 Quantity of teaching tablet computers owned by 10 thousand students in primary schools in various regions

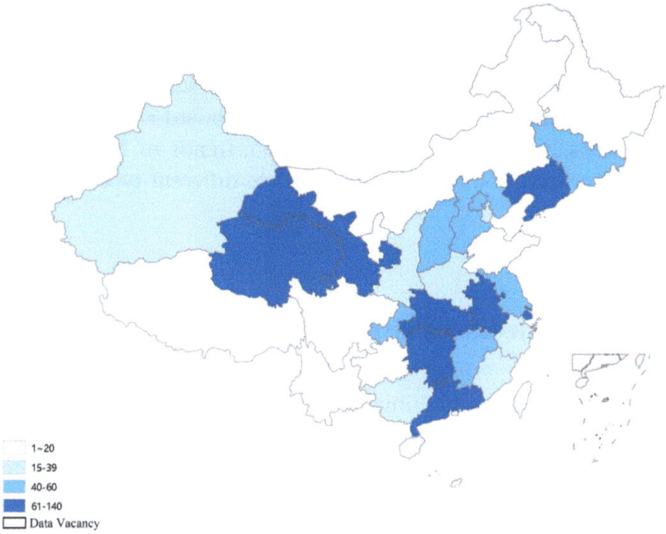

Province	Quantity
Inner Mongolia	1
Yunnan	1
Hainan	1
Ningxia	4
Shandong	10
Sichuan	11
Heilongjiang	11
Tibet	12
Guizhou	14
Shaanxi	14
Fujian	16
Xinjiang	18
Zhejiang	25
Henan	28
Guangxi	35
Tianjin	38
Shanxi	39
Chongqing	39
Jiangsu	42
Beijing	44
Jilin	46
Hebei	48
Jiangxi	54
Qinghai	60
Gansu	62
Anhui	62
Hunan	63
Guangdong	71
Shanghai	76
Hubei	104
Liaoning	131

Fig. 5.12 Quantity of teaching tablet computers owned by 10 thousand students in junior high schools in various regions

5.12.3 Tablet Computer Classrooms in Senior High School

The quantity of teaching tablet computers per 10 thousand students in junior high schools in various regions is shown as Fig. 5.13. In Beijing the quantity is relatively bigger, and in Inner Mongolia and Yunnan the quantity is relatively smaller.

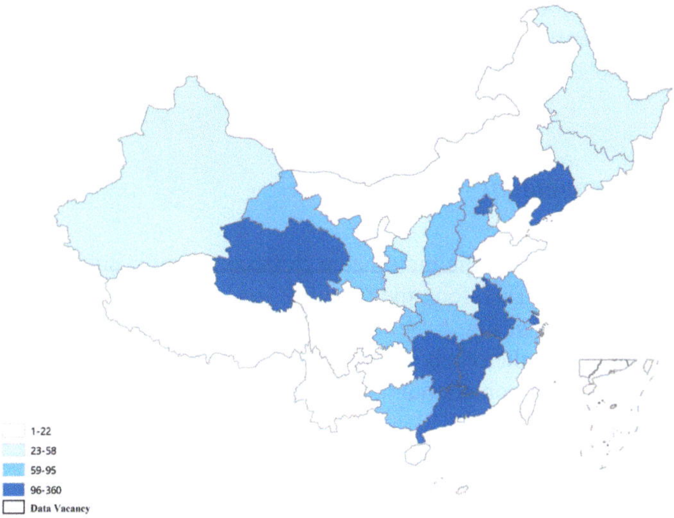

Province	Quantity
Hainan	1
Inner Mongolia	1
Yunnan	2
Ningxia	4
Sichuan	18
Shandong	20
Guizhou	20
Xinjiang	22
Jilin	28
Heilongjiang	37
Shaanxi	39
Henan	42
Fujian	43
Tianjin	47
Shanxi	59
Hebei	59
Zhejiang	63
Gansu	71
Guangxi	74
Jiangsu	82
Chongqing	86
Hubei	93
Jiangxi	96
Anhui	103
Guangdong	104
Qinghai	114
Liaoning	123
Hunan	125
Beijing	336
Shanghai	354

Fig. 5.13 Quantity of teaching tablet computers owned by 10 thousand students in junior high schools in various regions

5.13 Ways to Build Smart Classrooms in China

From the analysis of the status of classroom, we understand the status of multimedia classrooms, computer room classrooms, interactive white board classrooms and tablet computer classrooms in China. Combined with the trends of ICT in classroom and the status, we proposed some ways to upgrade different types of classroom environments.

5.14 Upgrading Multimedia Classroom

As a widely used classroom type, the multimedia classroom will be changed in the future. The upgrading of smart classrooms is analyzed below from four perspectives: classroom equipment (what mainly refers to the supply of hardware, spatial arrangement, teaching application (what refers to the method of integration between classroom equipment and teaching in new generation of classroom environment), and typical cases (what introduces the case of successful teaching in certain classrooms).

5.14.1 Classroom Equipment

Multimedia classroom is a type of classroom with relatively high coverage rate at present. This type of "high-definition display" classroom can well support the development of the "transmitting-accepting" teaching model. It is especially suitable for the basic teaching process of knowledge content concentrated teaching and it will be widely and generally utilized in the future. The equipment configuration situation of multimedia classrooms is shown as Table 5.1.

5.14.2 Spatial Arrangement

Viewing from technological aspect, the multimedia classroom mainly supports the instructive classroom teaching. Therefore, the "seeding" and "U-layout" classroom spatial arrangements as shown in Fig. 5.14 are commonly used. It has no definite limitation to the number of attendance, and sufficient multi-screen display as per the scale of the classroom, and can contain hundreds of people.

Table 5.1 The list of devices allocated for new generation of multimedia classroom

Devices	Traditional multimedia classroom	New generation of multimedia classroom
Projector + screen	✓	✓
Computer used for teaching purpose	✓	✓
Multimedia console table	✓	✓
High-speed wireless network		✓
Omnidirectional tracking high-definition camera		○
Multiple display screens		✓
Cloud service platform		✓
Intelligent censor		○
Open-source teaching platform		✓
Chairs that can be moved and assembled easily		✓
Interactive feedback device		○
Tablet PC used for teaching purpose		✓
Virtual simulation device		○

○ represents for optional device; ✓ represents necessary devices

Fig. 5.14 "Seeding" and "Multi-U-layout" multimedia classroom spatial arrangement

5.14.3 *Teaching Application*

The smart multimedia classroom environment is usually equipped with devices such as multi-screen display, interactive feedback equipment, tablet computer for teachers and virtual simulation equipment, which allows effectively integration of information technology, digital resource and classroom teaching. The featured teaching case in the classroom environment is shown as Table 5.2.

Table 5.2 Example of teaching application in smart multimedia classroom environment

Index	Teaching integration	Remarks
Resource access	Timely and convenient transmission	In virtue of high speed wireless network, teacher can acquire resources in a timely way
Content presentation	Multi-screen presentation	(1) Different formats of the same resource, for example, one screen for text and the other screen for animation (2) The 2nd piece of slide is automatically shown on the other screen when displaying the 2nd piece of slide
	"High-definition display" presentation	"High-definition display" to realize the immediate transmission of the content of tablet computer. For example, slide on the panel with fingers, and display high-definition video on the projection curtain to assistant to teaching
	3D presentation	It is available for teachers to realize the 3D presentation of image by 3D projector equipment
Learning activity	Voting	Evaluate the works and achievements of learning activity on the basis of interactive feedback equipment
	Responder	Develop teaching activity of knowledge competition on the basis of interactive feedback equipment
	Synchronous classroom	Realize special classroom by connecting with remote students by multi-screen
Learning interaction	Somatosensory learning	Conduct interaction with screen directly with body movement by utilizing somatosensory equipment
	Immediate evaluating	The students finish the online test, and the teachers can give immediate display and feedback of evaluating result

5.14.4 Typical Cases

The multimedia classrooms can effectively improve the construction of "Special Delivery Classroom", "Elite-Teacher Class" and "Elite School Network Classroom". Taking special delivery classroom as example, it can make the students in remote areas with insufficient teaching and research resources and weak teaching strength have the same class as the students of town center schools with abundant educational resources. For example, Enshi Tujia and Miao Autonomous Prefecture is situated in the poor mountainous area in southwest of Hubei Province; the rural schools are dispersed distributed in this prefecture, with relatively lagged school conditions, protruding classroom lacking problems and insufficient educational resources. Since 2010, this prefecture realizes "Village dream project", and enlarge and enrich the rural educational resources by utilizing multimedia classrooms, and form "consubstantial teaching" and "aid synchronous teaching" synchronous classrooms. The "consubstantial teaching" synchronous classroom shown

Fig. 5.15 Real picture of "Consubstantial Teaching" special delivery classroom of Erpo teaching point and Jinmao primary school

as Fig. 5.15 combines the classes of the same grade of the Erpo teaching point in mountainous area and the Jinmao Primary School, with lesson preparation, lesson giving, homework and exam at the same time.

5.15 Upgrading Computer Rooms

The computer rooms will be changed in the future. We will conduct analysis on aspects of the device allocation of traditional computer rooms and new generation of computer rooms, the available spatial arrangement, teaching application of new generation of computer rooms and typical cases.

5.15.1 Classroom Equipment

The Computer room is a type of special classroom environment. This type of "strong interactive" classroom can effectively support the acquisition of computer and simulation operation skills, which is also applicable to the research learning or project-based learning. The device configuration of computer rooms is shown as Table 5.3.

Table 5.3 Device configuration list for computer room

Device	Traditional computer room	New generation of computer room
Projector + curtain	✓	
Computer for teachers	✓	✓
Multimedia console	✓	✓
High speed wireless network		✓
Comprehensive tracking high-definition camera		○
Multi-screen		✓
Cloud service platform		✓
Intelligent sensor		○
Open source teaching platform		✓
Tables and chairs easy for movement and combination		✓
1:1 Computer for students	✓	✓
Exchange board	✓	✓
Server	✓	✓

5.15.2 *Spatial Arrangement*

The teaching model in a computer room environment takes the combination of demonstrative teaching and autonomous operation; the spatial arrangement is relatively single, and the "Queue" arrangement shown in Fig. 5.16 is the most commonly used. At the same time, with the occurrence of new types of teaching methods based on project-based learning and group association innovation, "Multigroup Round Table" computer room spatial arrangement is gradually becoming more popular.

Fig. 5.16 Spatial arrangement of "Queue" and "Multigroup Round Table" computer room

5.15.3 Teaching Application

The smart multimedia classroom environment is equipped with enough computers for students. Utilizing resource sharing platforms such as cloud service and online education, smart multimedia classrooms can effectively realize the in-depth integration between information technology, digital resource and classroom teaching, and the featured teaching case in the classroom environment is shown as Table 5.4.

5.15.4 Typical Cases

Classes developed by Hunan Changsha Vocational Institute of Civil Affairs focus on the practical operation and analog simulation. All the classes are developed in a computer room environment. Since September 2010, relying on the "vocational education new line" website, and using cloud computer network space as the platform in combination of high speed network computer classroom, developed the face to face teaching, and realized Connecting Students in Cyber Learning Spaces. All teachers and students of the school participate in the construction of teaching resources; teacher construction lessons as well as share teaching resources such as teaching plan and courseware in the spaces; students finish homework and collect data in spaces; teachers and students conduct interaction and communication in spaces, and share teaching and learning resources. Figure 5.17 shows that students are learning in computer rooms relying on cloud computer network space. Up until

Table 5.4 Example of teaching application in smart computer room environment

Index	Teaching integration	Remarks
Resource acquisition	Timely and convenient transmission	In virtue of high speed network, students acquire immediate resource from teacher end or the internet
	Connecting students in cyber learning spaces	Students acquire immediate resource from individual network space
Content presentation	Multi-screen presentation	Students present the works on non-main teaching screen
Learning activity	Group association learning	Group association to finish software operation, to design innovative works
	Remote group association	Develop remote online association and exploration on basis of terminal, to finish the learning tasks together
Learning interaction	Remote control	Teachers can conduct real time remote control and sharing by broadcasting
	Paperless exam	Give immediate feedback to paperless exam result, and conduct re-test aimingly

Fig. 5.17 Real picture of cloud computing network space learning in Hunan Changsha Vocational Institute of civil affairs

the end of 2014, the teaching resource has begun to take shape: more than 900 space resource lessons constitute the lesson wall, 12 network live system, 10 great master studios constructed by the school and enterprises, 3.2 million network teaching resources as well as 130 million network teaching resource page view.

5.16 Upgrading Interactive White Board Classroom

The interactive white board classrooms will be changed in the future. We will conduct analysis on aspects of the device allocation of traditional interactive white board classrooms and new generation of interactive white board classrooms, the available spatial arrangement, teaching application of new generation of interactive white board classrooms and typical cases.

5.16.1 Classroom Equipment

The interactive white board classroom has been popular in recent years. This type of "strong interactive" classrooms can well present the teaching contents in classroom teaching, and operations such as drawing and dragging can be conducted on the screen. The device configuration situation of interactive white board classrooms is shown as Table 5.5.

Table 5.5 Device configuration list for interactive white board classrooms

Device	Traditional interactive white board classroom	New generation of interactive white board classroom
Computers for teachers	✓	✓
Multimedia console	✓	✓
High speed wireless network		✓
Comprehensive tracking high-definition camera		✓
Multi-screen		✓
Cloud service platform		○
Intelligent sensor		✓
Open source teaching platform		✓
Tables and chairs easy for movement and combination		✓
Interactive feedback equipment		○
Interactive electric white board	✓	
Short focal projector	✓	
Large screen touch control liquid crystal all-in-one machine		✓
Large screen touch control interactive desktop system		✓

5.16.2 Spatial Arrangement

The interactive white board classrooms can efficaciously support classroom teaching and group association interactive learning method. Therefore, the "Seeding" spatial arrangement shown as Fig. 5.18 is commonly used. With the

Fig. 5.18 "Seeding" and "Multi Round Table" interactive white board classroom spatial arrangement

occurrence of new types of learning methods such as the group association innovation or the “Multigroup Round table” which lay stress on association is gradually popularized.

5.16.3 Teaching Application

The smart interactive white board classroom environment is equipped with large screen, touchable large-sized screen interactive desktop system, and tablet computer for teachers. It can effectively realize the in-depth integration of information technology, digital resource and classroom teaching. A teaching example of the classroom environment is shown as Table 5.6.

5.16.4 Typical Cases

The interactive electric white boards are popularized in the 42 teaching classes of Beijing Dengshikou Primary School, as shown in Fig. 5.19. Based on the

Table 5.6 Example of teaching application of smart interactive white board classroom environment

Index	Teaching integration	Remarks
Resource acquisition	Timely and convenient transmission	The group acquires resource from internet or teacher terminal based on interaction desktop system
	Convenient resource sharing	Real time sharing works between groups
Content presentation	“High-definition” presentation	Realize the presentation of group interaction desktop system in tablet computer for teachers
	Virtual simulation	Finish the experiments which are risky or with high cost in traditional experiment environment
Learning activity	Group association	The activity based on large desktop display screen, for example, association creation and association correction
	Synchronous classroom	Connect with remote students by multi-screen, to develop association learning
Learning interaction	Learning tracking	Record the teaching activity and test result during the interaction between students and equipment, and give evaluation report
	Content mark	Students give signs on key and difficult points as well as review to video resource based on after class review feedback equipment
	Interactive demonstration	Teachers develop interactive demonstration based on the touch control all-in-one machine, for example dragging, zoom and drawing

Fig. 5.19 Real picture of classroom teaching in future classroom of experimental primary school in Suzhou District Suzhou City

interactive teaching using the white board, it can change the thinking methods of students and improve their learning quality. By utilizing interactive white boards, and based on teaching subjects, teachers can make gradient design, and can write on the white board, and control computer programs to realize the effective synchronization of computer, projector and interactive white board. Teachers demonstrate any content in the computer for audience by the projector, and operate the computer by the interactive white board, without being in need of returning back to the computer. The white board is integrated in the daily classroom teaching, to make teachers more passionate and provide more interaction between students and teachers, as well as to make students have more autonomous learning.

This type of classroom mainly serves for expanding teaching tasks of Chinese, Mathematics and English, once a month for each class on average. The "one-person-one-machine" teaching environment in Future Classroom realizes the autonomous inquiry learning which takes "learning" as the principal thing, for example: on the Chinese composition class "Wonderful 3D Printing Technology", teachers make students have sufficient interaction, and conduct independent inquiry of 3D printing technology. During the teaching process, students can obtain more visual feelings by watching video 3D printing effect. According to the investigation, more than 90% teachers who have used future classroom express sufficient confirmation on the role in interactive, discussion and inquiry teaching.

References

Chu, S. L., Quek, F., Bhangaonkar, S., Ging, A. B., & Sridharamurthy, K. (2015). Making the maker: A means-to-an-ends approach to nurturing the maker mindset in elementary-aged children. *International Journal of Child-Computer Interaction, 5,* 11–19.

Duanyuan, B. A. I., Ji, Z. O. U., & Zhang, C. (2016). Application and effect analysis of task driven teaching mode in the course teaching of automation specialty. *Studies in Literature and Language, 11*(4), 13–17.

Erstad, O. (2010). Educating the digital generation. *Nordic Journal of Digital Literacy, 5*(01), 56–71.

European Commission. (2006). Recommendation 2006/962/EC of the European Parliament and of the Council of 18 December 2006 on key competences for lifelong learning. *L 394 Official Journal of the European*, 10–18. Retrieved from http://www.cmepius.si/files/cmepius/userfiles/grundtvig/gradivo/key_competencies_2006_en.pdf

Hedge, T. (1993). Key concepts in ELT: Fluency and project. *ELT Journal, 3,* 275–277. doi:10.1093/elt/47.3.27

Hong, K. S., & Songan, P. (2011). ICT in the changing landscape of higher education in Southeast Asia. *Australasian Journal of Educational Technology, 27*(8), 1276–1290.

Huang, R. H., Yang, J. F., Zhou, Y., & Gao, L. (2014). A study on learning preferences and class innovation of primary school. *E-education Research, 2,* 006.

Huang, R. H, Hu, Y. B, Yang, J. F, & Xiao, G. D. (2012). The concept and characteristics of smart classroom. *Open Education Research*, *2*, 22–27).

James, P., & Hopkinson, L. (2009). *Sustainable ICT in further and higher education*. Final version, 13.

Kinshuk, Chen, N. S., Cheng, I. L., & Chew, S. W. (2016). Evolution is not enough: Revolutionizing current learning environments to smart learning environments. *International Journal of Artificial Intelligence in Education, 26*(2), 561–581.

Lubold, N., Walker, E., & Pon-Barry, H. (2016). Effects of voice-adaptation and social dialogue on perceptions of a robotic learning companion. In *The Eleventh ACM/IEEE International Conference on Human Robot Interation* (pp. 255–262). IEEE Press.

MaKinster, J., Trautmann, N., & Barnett, M. (2014). *Teaching science and investigating environmental issues with geospatial technology*. Dordrecht, The Netherlands: Springer.

Martin, L., & Dixon, C. (2013). Youth conceptions of making and the Maker Movement. In *Interaction Design and Children Conference*. New York.

Mattsson, J. E., & Mutvei, A. (2015). How to teach evolution. *Procedia-Social and Behavioral Sciences, 167,* 170–177.

Miller, L. M., Chang, C.-I., Wang, S., Beier, M. E., & Klisch, Y. (2011). Learning and motivational impacts of multimedia science game. *Computers & Education, 57,* 1425–1433. doi:10.1016/j.compedu.2011.01.016

Moeller, B., & Reitzes, T. (2011). *Integrating technology with student-centered learning*. Quincy, MA: Education Development Center Inc, Nellie Mae Education Foundation.

Novosadova, M., Selen, G., Piskunowicz, A., Mousa, S. H. N., Suoheimo, S., Radinja, T., et al. (2007). The impact of non formal education on young people and society. In M. Nomikou (Ed.), *Non formal education book* (pp. 1–58). Belgium: AEGEE-Europe.

Palfrey, J., & Gasser, U. (2008). *Born digital: Understanding the first generation of digital natives*. New York: Basic Books.

Parker, C. E., Stylinski, C. D., Bonney, C. R., Schillaci, R., & Mcauliffe, C. (2015). Examining the quality of technology implementation in stem classrooms: demonstration of an evaluative framework. *Journal of Research on Technology in Education, 47*(2), 105–121.

Peeraer, J., & Van Petegem, P. (2015). Integration or transformation? Looking in the future of Information and Communication Technology in education in Vietnam. *Evaluation and program planning, 48,* 47–56.

Power, C. (2007). Educational research, policy and practice in an era of globalization. *Educational Research for Policy and Practice, 6*(2), 87–100.
Ringstaff, C., Sandholtz, J. H., & Dwyer, D. (1991, April). Trading places: When teachers utilize student expertise in technology-intensive classrooms. Paper presented at the annual meeting of the American Educational Research Association, Chicago.
Song, H., Kim, J., & Luo, W. (2016). Teacher-student relationship in online classes: A role of teacher self-disclosure. *Computers in Human Behavior, 54,* 436–443.
Tapscott, D. (1998). *Growing up digital: The rise of the net generation.* New York: McGraw-Hill.
Tapscott, D. (2008). *Grown up digital: How the net generation is changing your world HC.* McGraw-Hill.
Timms, M. J. (2016). Letting artificial intelligence in education out of the box: educational cobots and smart classrooms. *International Journal of Artificial Intelligence in Education*, *26*(2), 701–712.
Williams, P. (2005). Lessons from the future: ICT scenarios and the education of teachers. *Journal of Education for Teaching*, *31*(4), 319–339.

Chapter 6
Smart Learning in Corporate University

Abstract Corporate university is a platform providing further education for employees in the corporation. Smart learning advocated by the corporate university helps the corporation to maintain its uniqueness and succeed in fierce competition. This chapter introduces the development and current status of corporate universities in China, including their background, features, functions, teachers and the structure of courses. The connection of learning elements in the corporate university and smart learning is also identified with case analysis in this chapter. The online platform of corporate university is introduced by description of its framework, functional module, application and effective implementation. In the end, the future directions of corporate universities are discussed.

Keywords Corporate university · Smart learning · Learning elements · Online learning platform · Future development

6.1 Introduction

As described by Allen (2002), a corporate university is any educational entity that is a strategic tool designed to assist its parent organization in achieving its goals by conducting activities that foster individual and organizational learning and knowledge. According to Jeanne Meister, corporate university is the chief vehicle for disseminating an organization's culture and fostering the development of not only job skills, but also such core workplace skills as learning-to-learn, leadership, creative thinking, and problem solving (Meister 1998a). It develops because of "the emergence of the knowledge economy" (Meister 1998b).

Corporate university was also studied by Ramli (2008), who drew a conclusion that the following explanation aptly describes it, which is "A Corporate University (also known as an Academy, Institute, learning center or college) is an organizational entity dedicated to turning business led learning into action. It is designed, driven and intricately linked to the company's business strategy with the aim of achieving corporate excellence through improved staff performance and a

D. Liu et al., *Smart Learning in Smart Cities*, Lecture Notes
in Educational Technology, DOI 10.1007/978-981-10-4343-7_6

company-wide culture in which innovation can thrive. In addition to generating value from their intellectual assets, it helps organizations to identify, retain and promote key employees, whilst at the same time providing valuable, work based learning and career development opportunities for staff."

Corporate universities (CU) are a growing trend in corporations. In 1993, corporate universities existed in only 400 companies. By 2001, this number had increased to 2000, including Walt Disney, Boeing, and Motorola (Denise 2002). Just as Zeng and Liu (2006) stated, "The education of enterprise university research cooperation is the demand of economic and social development and is the demand of the enterprise and school development."

The concept of corporate university (CU) originated in the United States in about the early 1920s and became popular with most of organizations of America after the 1980s. (Ji 2011) However, CU in China is a recent phenomenon although it existed more than eight decades in the western world (Sham 2007). The development of corporate university in China is a result of "drastic economic growth and fundamental structural change" (Sham 2007). Just as Jiang (2012) said, university of China enterprise development in recent years has caused widespread concern in the various sectors of the community, began to rapidly increase in number, has become the fastest growing area of Chinese higher adult, the most striking part. And there is no doubt that CU as a new education model in China and will last 10 years also has made significant progress in development, to promote the development of knowledge economy and social progress provides a powerful intellectual support (Jiang 2012).

It is obvious that CU has been studied by many excellence experts for a long time, such as the comparative study of the traditional public university and the corporate university, which raise questions as to the appropriateness of the corporate bodies adopting the term "university" (Blass 2001). Paton et al. (2005) analyzed the initiatives reasons of the corporate university and Levinson (2007) interprets the main goals of corporate university. What's more, Anderson and Lichtenberger (2007) present an overview of the corporate university landscape in Germany. Dealtry (2000) sets the scene for the exploration of corporate and company developments in the field of the corporate university and related development paradigms. In addition, Prince and Stewart (2002) try to offer a contribution to enabling an understanding of the concept of the corporate university to be developed. Amy and Jessica (2012) explore and advance the understanding of the corporate university phenomenon. In a conclusion, most of researchers are focusing on the concept, the origin and development models of corporate university. However, the future development trend and learning methods of corporate university are rarely mentioned in the former studies, not to mention the new term smart learning.

In the Internet+ era with rapid development of information and technology, smart learning has become more and more important. Under this situation, it becomes unavoidable to popularize smart learning in corporate university. Smart learning is changing the way we learn and it is starting to change the way we live. This study is going to investigate the smart learning in corporate university and the relationship between them. For the smart learning in corporate university has become an inevitable trend in the future on account of its intelligence and adaptability, it should be the issues drawing the close attention of researchers. Analyzing the corporate university from the perspective of smart learning may stimulate corporate universities to rethink their future development tendency as well as management direction.

6.2 History of Development and Current Status of Corporate Universities

The enterprise is one of the important places for staffs' learning. When the era of knowledge economy is globally coming, the enterprise competition will be finally ascribed to the competition of talents, and will be further embodied as the competition of staff quality and staff learning ability. Therefore, it is the important contribution of the corporate development strategy in this period to acquire and foster excellent staffs and management talents, and integrate management staffs with corporate culture. Corporate university is undoubtedly one of the main pathways to complete development strategy.

6.2.1 Developmental Background of Corporate Universities

In the 1920s, the degree of corporate intervention in education and occupational training continued to increase, and the development of high-qualified human resource has become the core element to improve corporate comprehensive competitiveness. In this condition, corporate university emerges as a new corporate education mode. Three main factors account for the emergence of corporate universities as follow:

1. Responding more effectively to the global challenges. The increasingly intensive competition environment proposes new requirements for corporate development: an ability to integrate learning strategy with commercial target; an ability to increase learning opportunities for staffs; etc.
2. Establishing a new form of continuing education. Corporate university emphasizes on common development between organization and individual, and provides for people in various ages with multiple educational opportunities in convenient and timely ways.
3. Supporting the staffs' occupational demand. New contractual relationship is formed between the corporate and the staff, that is, the corporate provides for staffs with opportunities to enhance abilities, and staffs will pay back their higher production capacity and commitment to corporate mission during the working period.

6.2.2 Features and Functions of Corporate Universities

Corporate learning regards the whole corporate as a learning subject (Ke 2010), which is a series of management activities to improve the staff quality, ability and job performance, and its core is to learn and gain knowledge. Corporate endows staff with the opportunity of learning while working, and utilizes technology to achieve the maximization of learning and outcome. The effectiveness of corporate learning depends on the attitude of enterprise manager. It is necessary for enterprise manager to consider the value and significance of learning based on decision-making level, and formulate the corporate learning strategy in accordance with the requirements of the corporate and the staffs. Meanwhile, corporate university, built on a relative mature stage of corporate development, is regarded as an organization supplying learning service. It embodies the escalation of administering functional department, and represents the spread and extension of service client. (Please refer to Table 6.1).

The word "Corporate University" was firstly adopted by Walt Disney in 1950s. Corporate university is also known as Company University, which is a new type of educational training organization funding by the corporate, and propel the learning of individuals and organizations by various planned learning activities, and propel knowledge and intelligence development, to support the corporate in carrying its mission (Li and Wu 2013). Corporate university is a senior form of corporate education development, and one of the important means for the corporate to effectively improve core competence and maintain the uniqueness.

Corporate university differentiated in its orientation, teachers and training methods as following (Wu and Liu 2013):

- It is established by corporate funds, and it owns the self-development goal and strategy;
- The internal trainers lead all business lines, and also are assisted by the teachers from higher education institutes, research institutes and professional training institutes;
- The training method is based on analogue simulation and mobile platforms to implement self-course design, interactive self-help;
- Form the educational and training system which integrates the value chain process externally, and integrates tangible business and staff ability promotion internally;
- It does not aim to education and training, but it pays more attention to practical problems and the discovery of human resources in the enterprise.

The functions are mainly realized by corporate university are (Jiang 2012):

- To provide a guarantee for corporate strategy and corporate development. As a part of corporate development strategy system, corporate university supports a series of reserve force supply and related resource expanding work.
- To provide an effective path for corporate cultural construction. By the effective utilization of internal and external training paths, corporate university offers help

Table 6.1 Development stages and features of corporate learning

Development stage	Initiate stage		Growth stage	Mature stage		Sustainable development
Learning service organization evolution	• Administration department • Personnel department	• Human resource department • Training department	• Training center • Learning development center • Human resource development center	Inward corporate university	External corporate university in service of supply chain	Export-oriented corporate university in service of society
Corresponding features	• Emergency training without professional training post or definite training plan • Transit to human resource department gradually	• Establish professional training post, and starts to plan for annual training plan • Select courses as per organization development and post demands, and starts to establish training system • Establish training system and starts Training Department, which belongs to Human Resource Department	• Established relatively perfect training system, and starts to pay attention to staff occupational career plan, and formulates corporate internal training courses, and establishes teacher system • Training institute independent with Human Resource Department, and is the early form of corporate university	• On hardware aspect, it has independent training base and advanced teaching facility • On software, it has perfect training system and management system	The service objects are expanded to supplier and partners, and can support supply chain integrity and business development	• It has advanced training resource and brand influence power. • The service objects are expended to the whole society, and can promote corporate image and benefit
Service object	Middle-layer staffs in the corporate	Middle and high layer staffs in the corporate	Staffs in the corporate	The whole supply chain		The whole society

for corporate culture construction. The training of corporate university to external students, potential customers and strategic partners can make them understand the corporate culture better, and solidify the loyalty and stable relation between cooperators.

- To provide featured education service for corporate development. By carrying out various education activities, corporate university can specifically and efficiently provide high quality education service for different level staffs in the corporate, in order to cultivate excellent knowledge-based staffs and to meet the requirement of high quality human resource development.
- To optimize knowledge resource in order to realize the long term development of the corporation. As the upgraded version of traditional corporate educational institute, the corporate university is just like a resource integrator, which can capture the forefront knowledge resource of the corporate, to concentrate various types of knowledge resources into the corporate university to a maximum extent.
- To foster high quality staffs for the corporate. Corporate university is not only the complementation to knowledge and training to skills of staffs, it injects new vitality to the corporate by the potential ability development to staffs.

6.2.3 Teachers and Course System in Corporate University

Based on corporate practice activities, the Corporate University is closely related to government policies, technological reform and cultural concept; in order to be "innovative, practical and effective", it is necessary to change the learning content of corporate, to make it follow the change of times, and the teachers of corporate university present flexible and diversified features.

The faculty contains external teachers and internal teachers:

1. External teachers engagement. The external teachers mainly include professors of higher learning institutions and lecturers of professional management consulting companies. They are equipped with advanced management philosophy and thoughts, abundant teaching experience and diversified teaching skills. However, they are usually the experts in certain course field; they give almost the same lessons when faced with different industries and corporate with different requirements, therefore, they will not provide substantial significance and help for the practical operation of the corporate.
2. Internal teachers selection. The channels for corporate university to select internal teachers include: middle and high layer management staffs of the corporate, middle and high layers of technicians, business backbone as well as grass-roots staffs with excellent achievement or extruding contribution. The superiority (advantages) embody on that: the senior managers in the corporate clarify the training requirements and development objectives of the corporate, and the

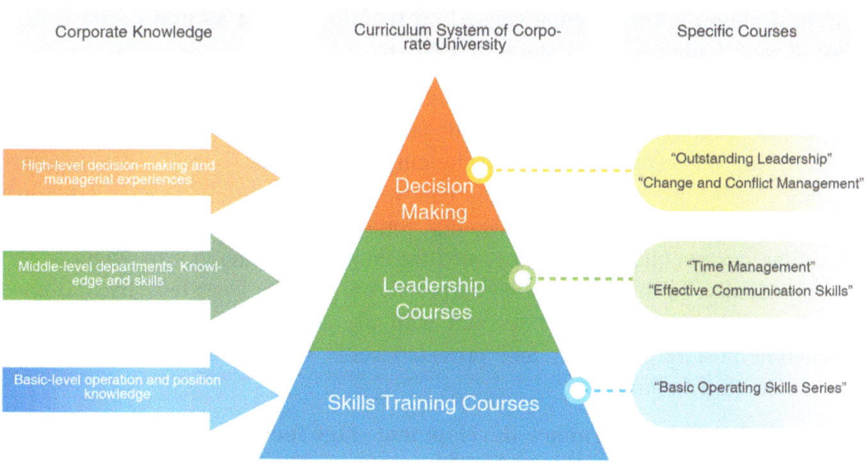

Fig. 6.1 Corresponding relation between corporate knowledge and corporate university curriculum system

course design and teaching content are more practical and targeted; the teachers are senior corporate managers, which makes it easier for the broadcasting of strategic objectives and corporate value system, corporate culture and corporate vision. But relatively, they have limited teaching skills and cannot effectively motivate students; they have relatively less teaching experience (Tian 2013).

The course system of the corporate university is composed by unique knowledge of different positions and the public knowledge of the whole corporate. They form the three-dimensional, complicated and internally disciplinary organization structure including corporate staff layer structure, corporate and staff development stage. The course objective of corporate university can be analyzed, developed and concluded as per the specific work responsibilities of different posts, or the different knowledge needed by different staff. Corporate university courses can be divided into three types: the first one is general type, mainly refers to culture, occupational quality and basic capacity; the second one is specialty/technology type, mainly refers to marketing, audit, finance, human resource, operation, market and public affairs as well as professional basic skill; the third one is management class, mainly refers to management basic skill, management move forward and leadership promotion (Fig. 6.1).

6.2.4 Revolution of Corporate Universities in China

The corporate university in our country appeared and developed with the change of corporate development and competition environment, which can be divided into the following three development stages:

The first stage: embryo stage regarding training school as main part (beginning of new China—initial stage of 1990s).

In the planned economy times in the last century, many corporates established their own corporate technical schools or staffs and workers universities to cultivate their staffs; taking academic education as the emphasis, it became a kind of important supplementation to national academic education. In the 1980s, with the establishment of the market economy, the idea that corporate training school takes academic education was not changed, but it turned the emphasis on the improvement of staff skills gradually. In this stage, most corporate universities were held by traditional manufacturing industry corporate, and with the aim of satisfying the corporate production post requirement, and provided staffs with immediate and practical production skill training.

The second stage: the primary development stage focusing on improvement of staff skills (early 1990s—later 1990s).

In the early 1990s, with the intensive corporate competitions, there were abundant amount of overseas funded corporate rushing into China, and a large number of private corporate came up. In 1993, Motorola China Zone Corporate University established. Under the driving of foreign-funded enterprises, many domestic corporates began to realize the superiority and importance of learning in cultivation and improvement of corporate competitiveness. During this period, the corporate training schools in our country began to change into the corporate university taking improvement of staff skills as the emphasis, with their main function as a training center and relatively low training layer. The training contents mainly include post requirement and individual performance improvement. Most of the training teachers are external teachers, and most students are internal staffs of the corporate.

At this stage, emphasis is laid on training for the corporate internal staff. It is the most primary and direct purpose to conduct systematic training of the internal staffs to improve their comprehensive quality and professional ability.

The third stage, the reform period taking realization of corporate strategy as the emphasis (later 1990s to today).

In the late 1990s, the development of corporate university was in a management reform period focusing on improving competitiveness. Corporates required their universities to improve corporate competitiveness and promote organizational reform, to realize the corporate strategy objective. According to the strategic requirements, the objective for corporate university training and development may be extended to the members of the upstream and downstream of the whole commercial value chain, and it may provide staffs with educational background learning, such as MBA.

In this stage, by providing customers, staffs, suppliers, partners and other potential customers with targeted service training project, to realize the win-win on each process of the industry chain.

Corporate universities in the past usually laid stress on standardized construction, as well as training systems, teaching team and course development, the work mode of which can be call as fragmented and processed standardization construction, which makes corporate university input a large quantity of resources and energy, but obtain limited effectiveness. With the evolution of university development process, different from traditional standardized thought, the modern corporate universities lay stress on result-oriented thought; it is necessary to utilize systematic thought and grasp the critical process of corporate university value creation, and conduct effective resource configuration, to create the maximum value (Table 6.2).

Specific development situations of corporate universities in our country:

The corporate universities originated from the Crotonville college established by General Electric Company (GE) in 1956. In 1993, Motorola firstly introduced the concept of Corporate University into China; in 1998, Hisense and Chunlan established their own corporate universities, respectively; after 2008, the number of corporate universities increased in a speed of 236 each year and 1.5 each day. Up until the end of 2012, the number of corporate universities in China had reached 1186. And then, the number was increased by 800 each year. Up until 2014, the corporate universities in China had been more than 2500. And nearly 50 million people accept the education of corporate universities (Fig. 6.2).

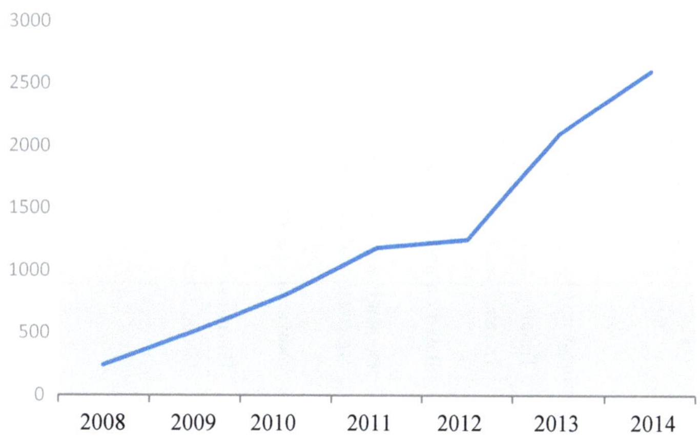

Fig. 6.2 Corporate university establishment quantity trend during 2008–2014

Table 6.2 Development revolution of corporate universities

	Training item	Learning item	Performance item
Leading personnel	Training teachers	Students + instructors	Students
Training mode	Based on ability + task	Based on ability + task + problem	Based on ability + task + problem + performance
Roles of training department	Demand analysis, purchase instructor and organizational implementation	Analyzing and planning, purchase instructor and implementation together	Providing method and tools, and integrating expert resources
Roles of students	Individual learning application	Solving problems for group learning	Improving group performance
Roles of experts	Formulating course and giving lessons	Planning, giving lectures and coaching	Giving lessons and coaching
Project effectiveness	Group ability improvement	Organization ability promotion	Organization performance promotion

6.2.5 Smart Learning in Corporate University

According to the global environment uncertainty, it is required for corporates to conduct continuous updating and recycling based on a learning organization to obtain success. As for corporates, learning becomes a kind of basic competition element and survival method, and all the changes propose new requirements to the organization type and learning ability to the corporate. Smart learning in corporate university is one of the formats for evolution of learning organization. It means that organizations have the ability of continuous learning, and can give full play to the innovative thoughts of staff. It is a kind of organizational atmosphere which can form learning sharing and interaction. The smart learning in a corporate university becomes the inevitable trend of corporate development because of the intelligence and adaptability.

Corporate digital learning is a process in the corporate or between corporates that promote group interaction and knowledge sharing through information and communication technology. To realize the application, conversion and creation of corporate knowledge, to improve corporate adaptability and competitiveness (Ke 2010). This is also the obvious features of smart learning in corporate university. This form of learning focuses on corporate staffs or corporate learners, and is a new learning model with complete learning experience.

Features of smart learning in corporate university:

1. Information technology as the main support.

The smart learning in corporate university must be established on the basis of effective informationized learning environment and mechanism, to realize the information acquisition, broadcasting, explaining, sharing and knowledge conversion of learning subject at any time and place, by any method and any step.

2. Offering diversified collaborative communication work.

The smart learning in corporate university can provide diversified collaborative communication tools, in convenience for corporate members to construct effective network practice community, to build up relaxed and efficient learning atmosphere.

The smart learning in corporate university includes four aspects of contents: staff learning activity, training teaching activity, staff learning content and corporate learning space-time.

Staff learning activity is the learning tasks, the learning method adopted and the learning result after finishing the learning objective of corporate staffs on learning,

Training teaching activity refers to the teaching strategy during selecting, organizing, adjusting and controlling of student learning activities by training institute, the learning support during learning activities, and the learning evaluation during and after learning activities.

Staff learning content refers to the learning media and learning resource of learning activities, as well as the learning objective when guiding students to develop learning activities.

Table 6.3 Instructions to features of smart learning of corporate university

Characteristic dimension of smart learning in corporate university		The learning condition should be achieved in corporate university in the future
Technology support	Tracking and analysis technology	Utilize learning behavior tracking technology, to obtain the learning behavior data of staffs, for example, learning system log in times, online time and public comment, to know the learning status of stuffs and give suitable and timely help
	Evaluation and support technology	On teaching support aspect, there will be digital teaching materials in the future, which will play the role of supplementation to paper teaching materials by virtue of abundant learning resource and learning activity management system
	Perception and adaptive technique	The interaction-based brand new informationized construction idea, in combination of whole set of teaching terminal equipment such as Smartphone, tablet computer and electric white board, and in integration of excellent and abundant teaching resources, to form man-machine interactive and instant interaction network learning environment
	Organization and reconstruction technique	The high technology means are used to enrich learning methods, for example, utilizing disk to conduct man-machine dialogue and self-coaching training, and utilizing terminal technology network to conduct large scale remote training, to achieve qualitative changes on training and teaching methods
Learning system	Staff learning content	The fragmented learning will become the mainstream learning form in the future. The learning of corporate university shall be in accordance with this feature, to achieve short length, focused content, to meet the requirements of querying when needed and using after learning of the corporate staffs in the future. The learning content can be extended to cases, experience sharing and some standard operation norms
	Staff learning activity	Provide corporate staffs with the double learning resource including face-to-face teaching and network teaching, which supports for not only collective learning but also individualized learning
	Corporate learning space-time	By technologies such as virtual classroom, immediate chatting and micro world, corporate staffs conduct learning, and extend corporate learning into the social life, but not limited to the corporate environment
	Training teaching activity	By means of group discussion and online chatting room questions and answers, the training teaching activities eliminate the boring and uneasy emotions during training process. By utilizing smart teaching environment, the classroom face to face teaching can grasp the requirements of staffs, in convenience for training instructors to adjust their teaching contents flexibly

Corporate learning space-time is the learning sequence for the corporate learning resource arrangement, the learning space of developing learning activity (including the learning space of physical learning environment and virtual learning environment) and the learning community of participating learning and discussing learning (Table 6.3).

6.3 Learning Elements in Smart Learning of Corporate University

At present, most of the corporate universities in our country are still in primary development stage. Although the number of corporate universities in our country is considerable, only a small part of them is in accordance with the connotation elements of corporate universities. With the development of the internet, and with the propelling of new generations of information technology such as cloud computing, big data and ubiquitous network, corporate universities are constructing smart learning environment of smart learning and smart education, and the frame of smart learning has been preliminarily formed.

6.3.1 Learning Elements in Corporate University

We will conduct a brief analysis on the learning elements of corporate universities in combination of smart learning frame, on four aspects of staff learning activity, training teaching activity, corporate learning space-time and staff learning contents.

1. **Staff learning activity.**

Staff learning activity includes elements of learning outcomes, learning tasks and learning methods. "Staff-centered" learning mode is the leading mode of smart learning, which can meet the individual, ubiquitous, immigrating and group intellectualized development requirements of learners (Feng et al. 2013).

In which individuality means teaching students in accordance of their aptitudes and making the best use of the circumstance aiming at learning activity. Individuality usually includes interface individuality and service individuality. For example, in the same course, learner A is provided with video explanation, and learner B is provided with character explanation. The ubiquitous learning means to provide learners with necessary, sufficient and well-organized learning resource at any place and time.

Learning in context is efficient, and it is necessary for smart learning to provide learners with necessary support for situated learning; this design shall be person-first and situation-based, and the learning situation can be perceived and evolved, to establish knowledge base according learning situation, to provide optimum learning service. Group intellectualized learning refers to a group composed by learners, teachers, experts and instructors, who often conduct communication during learning process, and share various learning resources and finish certain learning tasks together.

2. **Training teaching activities**.

Training teaching activities include elements of teaching strategy, learning support and learning evaluation. In the smart corporate training teaching activities, the teaching method lays stress on activity design, and it evaluates learning outcomes based on the adaptability of the cognitive characteristics of learners. In corporate universities, training teachers can utilize augmented reality technique to present various real learning situations, to make learners experience the learning objective in an immersive way, to enhance their learning interest and motivation. According to the learner preview result recorded by the system, the training teachers lay stress on the explanation of the difficult knowledge points for students; by utilizing the abundant learning resources provided by the system, they design various learning activities.

3. **Corporate learning space-time**.

Corporate learning space-time includes elements of learning time sequence, learning space and learning community. Training teachers can utilize the videos recorded by classroom recording system and the student generating resources recorded by learning support platform to conduct analysis, and rethink their experience and shortcomings during teaching process, and compile reflection and exchange with colleagues, or they can utilize the learning support platform to give coaching and communication to the problems proposed about homework.

The smart corporate learning space-time is the high end form of digital learning environment; it is the appeal of corporate staffs to learning environment development under social informatization background, and is also the support condition for effective promotion of teaching method evolution. Smart learning space-time aims to make the learning field percept learning situation, to recognize the characteristics of learners, to provide suitable learning resource and convenient interaction tools and record learning process and evaluate learning outcomes automatically, to improve the effective learning of learners.

It is required for the smart corporate learning space-time to promote the learning and development of learners, to make learners achieve happy and efficient learning in classroom. In order to achieve this aim, it is necessary to take learning environment and activity into consideration, in which environment includes physical environment and learning support platform. On physical environment aspect, it is necessary to provide high interactive teaching and learning equipment for classroom subjects, to effectively support for the acquisition, treatment and presentation of classroom subjects to learning resources. The smart environment control can provide preferable external environment for classroom subjects, and conduct adjustment on light, temperature, sound, background music and air quality according to the real-time state of classroom. Innovative space layout mainly takes the supply of more humanized tables, chairs and facilities for learners, and the implementation of combination of tables and chairs according to the requirement of teaching and learning activities, to form learning groups to be convenient for the development of group learning activities (Chen et al. 2012).

4. **Staff learning contents**.

Staff learning contents include elements of learning media, learning resources and learning objectives. Smart corporate university learning can conduct intelligent analysis on the target information submitted by learners, and help learners to acquire the needed resources in a quick and accurate way, and intelligently push related resources. Besides, learners can select learning resource from resource supermarket according to their learning requirements, and teachers can give suitable guidance during the selection process. The system pushed and self-selected learning resources will be put into the smart individual learning spaces for learners to conduct further filtering, screening and recombination of resources (Guo et al. 2014).

6.3.2 *Case Analysis on Corporate University Learning Elements*

1. **Overview of W university**

Founded in July 2007, W University is the talent cultivation base for a leading online game as well as a mobile internet platform developer and operator company in China. As the powerful supporter and important constitute part of the integrated strategy of W Company, W University concentrates on providing professional and systematic training service for corporate staffs, undertaking the responsibilities of inheriting corporate culture and promoting corporate reform. By abundant teaching methods, W University brings more experience and value for staffs, high quality customers, partners and other potential customers in a highly effective way, to deliver corporate culture and establish wider commercial ecological system.

W University is equipped with three functional colleges of Commercial College, Game College and Experience College as well as E-learning development group.

- Commercial College: serious training projects of leadership for middle and high level cadres, various management training and new staff professional training, to improve the occupational quality of management team and staffs in a long term and systematic way.

- Game College: concentrate on the cultivation of product creation and operation talents, and establish and perfect professional technological course and teachers system based on the corporate qualification system, and propel skill certification for key posts based on E-platform.
- Experience College: according to the requirement of the construction of various teams of the corporate, utilize experiential learning method to associate for the integration of new staffs and groups, to improve group work efficiency, and broadcast corporate cultural value system.
- E-learning development group: establish online learning and online authentication platform according to various training projects and courses, to make learners conduct autonomous learning.

2. Comments on smart learning elements in W university

W University has distinctive features on learning activity and learning contents:

Gamification learning is adopted on learning activity aspect. For example, on corporate culture promotion activity aspect, W University develops "Culture Regiment Campaign" and "corporate culture authentication test" activities. According to seven cultural value systems, and by means of answering questions and making breakthrough and dragon ball collection, it successfully implants corporate culture into all the staffs. The gamification learning method combines enjoyment with effectiveness, to achieve the purpose of popularizing corporate culture.

The micro lesson design methodology is adopted on learning content aspect. In coordination with the educational technology research of the corporate, it quickly organizes internal experiment and verification in the organization, and integrates internal and external resources, by researching micro lessons in the industry and generally analyzing real situation of the corporation, outputs the micro lesson design methodology in accordance with the corporate value system and departments, i.e., the "walking into micro lessons—disassembly of original lessons—design architecture—compiling script—courseware making" five steps from the concept and definition of micro lessons to the output of micro lessons.

6.4 Smart Learning in Corporate University Online Platform

In the knowledge economy times with the basis of globalization, informatization and network, the development and management of talents is one of the key elements for the corporate to win in the competition. The improvement of staff ability by corporate university online platform is the strategic selection for the corporate to maintain competitive advantage.

6.4.1 Framework of Corporate University Online Platform

The corporate university online platform is an organic, complicated and diversified functional entity composed by learners, instructors, learning environment and learning resource. Different from electric training or remote education, corporate learning system includes the mutual communication, as well as the establishment, management and sharing of knowledge between staffs, and it provides solution for the delivery, evaluation and management of the whole learning experience. The materialized presentation of corporate university online platform is establishing a network platform for learning and communication, to conduct effective training and evaluation for the staffs by this platform, to provide the corporate with a preferable environment for staffs to promote learning ability.

6.4.2 Functional Module of Corporation University Online Platform

At present, the corporate online learning system is realized mainly by three aspects of technology, content and service. The technology mainly refers to learning management system. The content mainly refers to training course material. As for the corporate, there are mainly new staff training, post skill training, management training, tool training and other general training. Service mainly refers to the overall plan and implementation to training system, course system and technology system.

The functions of corporate learning system are mainly embodied on the support of three groups: training managers, training teachers and corporate learners (Table 6.4).

Table 6.4 Corporate learning system modules and the functions

Corporate online learning system	Corporate learning system module	Functions and objectives
Technology	Learning management module	Learning plan, post learning, technology training, exam center, individual learning, questionnaire and voting
	Evaluation and learning tracking module	Learning statistical analysis, post learning tracking, training management statistic, occupational technology evaluation and resource statistic analysis
Content	Knowledge base management module	Platform course resource base and courseware mall resource base
Service	System management module	User management, post system management, role authority management, training teacher management and department and division institute management

6.4.3 Case Analysis on Corporate University Online Platform

The construction of Z University learning platform started in 2001, which was one of the first corporate to introduce remote training in our country. During the 14 years, it experienced the preliminary construction period of 2001–2007, the development period of 2008–2010 and the deepening stage of 2011 until today; with large scale investment and practice, continuous optimization and construction, the construction and operation effectiveness of remote training system has become an important business of the corporate and takes the leading position (Fig. 6.3).

1. **Online learning management system**

Z University creatively applies mature IT technology into the development of Group Level Learning Management System (GL-LMS), and it expands the narrow online learning to the whole process including online and face to face training and learning, which effectively supports the training and learning activities of the corporate, and promotes staff learning experience, active learning and self-management enthusiasm. The system mainly includes the following functions:

(1) Core learning functions; including requirement management, activity management, strategy management and content management.

- Requirement management, i.e., in combination of the change of various strategic objective of the corporate, and by organization, post, ability and technique system to identify and design learning requirement, to make staff learning objective matched with technology objective.
- Activity management means divide course learning into formal and informal learning activities, and automatically record the design and development, authority distribution, learning implementation and tracking feedback of each learning activity according to the property of the learning activity to be managed.

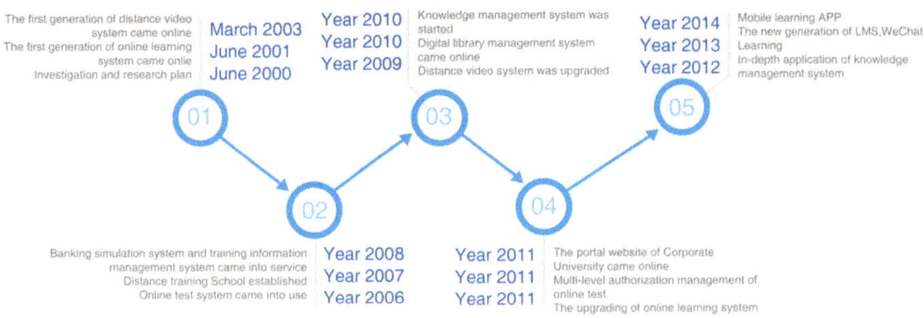

Fig. 6.3 Construction history of Z corporate university cloud learning platform

- Strategy management means formulating learning strategies in a flexible way, in support of combined learning.
- Content management means to adopt methods of learning community and conduct content management and information communication to all the staffs, and utilize a set of tool and management mechanism to convert the contents of information or knowledge layer into learning contents.

(2) Extended learning functions. Including the objective management, organizational management and performance management involved in learning, as well as staff occupational career plan.

- Objective management means to decompose the organizational objective into department, group and individual objectives, to automatically connect organizational objective with individual learning objective, to better support the realization and update of various levels of strategy objective.
- Organizational management can fully meet the learning requirements of the whole corporate, and various levels of managers can deploy and adjust the learning framework of the group.
- The performance management is butt jointed with performance management system, to analyze the performance data contents and formulate learning requirement and develop learning contents, and put feedback of learning performance into the performance system, as the data for performance evaluation.
- Occupational development plan means to draw the clear occupational development path by a series of data analysis, to associate the development of the organization and individuals, to provide matched learning resources and learning activities for each possible post promotion in the future.

(3) Report functions. Between core functions and extended functions, quantify management learning activities and provide visible statistic information service for managers and staffs.

2. **Online exam system**

It is developed by utilizing advanced cloud computing technology, with modularized, componentized and cross-platform architectural design, integrating functional modules such as information issuing, FAQ, personnel management, applying management, database management, exam paper management, exam management, questionnaire management, authority management and system management, fully meeting the complicated requirements of exam, competition, evaluation, investigation and analysis for corporations. In support of 5000 people online exam at the same time, and ensure the realization of large scale and high difficulty applications. Besides, the system can also conduct whole course dynamic supervision to the test process, to avoid cheating and form individual skill analysis report image.

3. **Knowledge management system**

Z corporate university takes the two dimensional viewpoint of knowledge-management and IT-governance, and starts from the top layer design, and concentrates on solving the long term knowledge management problems in the corporate; the developed knowledge management systems include public knowledge base, professional knowledge base, knowledge expert, knowledge questions and answers, knowledge engine, knowledge classification and knowledge map. The public knowledge base is mainly planned as a whole by the corporate university, and cooperated by various business lines; the construction of professional knowledge base is mainly dominated by business department, and guided and cooperated by corporate university; the knowledge management platform is unified planned by corporate university, and serves for various business lines.

4. **Simulated production system**

The so called simulated production systems means to establish a system independent from the real production environment of the corporate, and provide simulative practical operation service for new staffs, new institutions and new business. This system is equipped with most functions of the real production environment, but is totally isolated with the production environment physically, which is safe without operation risks.

5. **Digital library**

Digital library means the library utilizing digital technique to process and store various data. It stores various information resources of different carriers and different geographic locations with digital technique, which is convenient for the cross-regional object-oriented network query and broadcasting. It involves in the total process of information resource processing, storing, achieving, transfer and utilization. Popularly, digital library is a virtual library without walls, and it the extensible knowledge network system constructed and shared based on network environment, and it is also the extra large scale, distributed type, easy for using and space-time free knowledge center which can achieve cross-base seamless link and intelligent retrieval.

The construction of Z corporate university digital library can effectively promote the learning initiative, and is beneficial for the promotion of the comprehensive quality of staffs, and is the fundamental construction for the corporate to build up learning organization, and is the effective supplementation of training methods.

6. **Remote video system**

Remote video system means the system equipment used for individuals or groups of two or more places transfer the sound, image and document data by transmission line and multimedia equipment, to achieve immediate and interactive communication, to finish meeting or training objectives. This system is a typical image communication.

At the sending end of the communication, the image and sound signals are changed into digital signals, and they are reproduced into visible or audial information at the receiving end; compared with teleconference, it is more direct with larger information amount. The remote video system can make people listen to the voice as well as see the participants of the meeting, to make every participant feel as on the scene.

The Z corporate university remote video system is distributed all around various places, realizing the meeting training mode of combination of concentration and dispersion; the video is clear, and it is available for two-way communication and interaction, which improves the meeting training efficiency.

7. **Mobile learning system**

Mobile learning (M-Learning) means the remote learning by utilizing modern communication terminal such as mobile phone and PDA equipment (the laptops with wireless internet access function are generally not included) under the guiding of lifelong learning thought.

The mobile learning system of Z corporate university is mainly composed by Wechat terminal and App terminal. Mobile learning system is based on the cloud service solution, and the method of taking App client end as the emphasis and public Wechat as subsidiary is adopted, to realize the interactive popularity, to achieve the infinite sharing of learning resource, to gradually establish the "learner-centered and comprehensive" new learning platform.

As one of the corporate mobile learning gateways, the mobile learning App provides learning contents including training information, learning data, electric books and mobile courseware, and is the center for mobile learning of all the staffs. In the future, this system will open the level-to-level management authority, for various branch institutions (lines) to upload and download learning contents at the same time, and conduct butt joint to platforms such as knowledge management system, digital library and Yixuntong, to help staffs to realize the "one-stop" learning experience on the mobile end.

8. **implementation effectiveness of online platform**

In recent years, Z corporate university has obtained abundant implementation achievements in system learning field, which further releases the potential value of training and learning, and promotes the reputation and image of the corporate as well as the research influence of the university. The implementation achievement of the corporate learning system is mainly embodied as:

(1) Provide convenient online learning service experience for staffs, and effectively relieve the conflict between work and learning.

The learning management system continuously optimizes and promotes the training and learning efficiency, and realizes the unity of training and learning activity, and lays foundation for the construction of group learning ecology. In latest 5 years, the organization has developed 499 standardized electronic course wares, in which 219

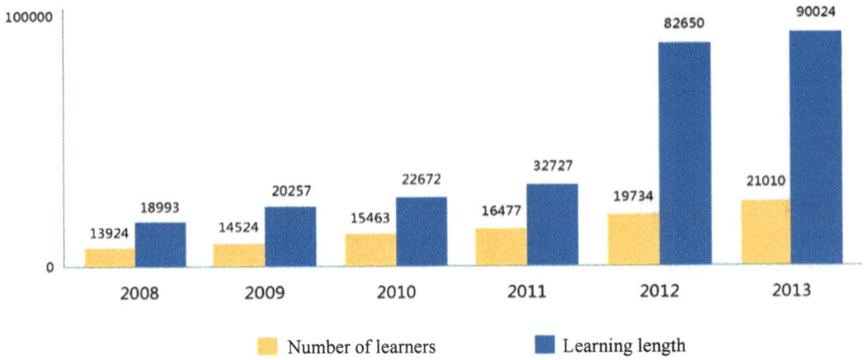

Fig. 6.4 Distribution of number of learners (persons) and learning length (hours) during 2008–2013

are self-researched and developed, and the proportion for self-research and development rises to 55% of 2013 from 30% of 2008; the online learning total length is 240,000 h. The substitution rate of electric learning to traditional learning rises to 30% of 2013 from 8% of 2008 (Fig. 6.4).

(2) Provide high-efficiency online exam and investigation evaluation service for all the staffs.

The online exam system has become the important platform for corporate post qualification exam and professional sequence exam. Up until now, there are 300 exam managers, with the capacity of 5000 people, and 2000 exams are organized each year, with total number of people who participate in exams of 200,000 (Fig. 6.5).

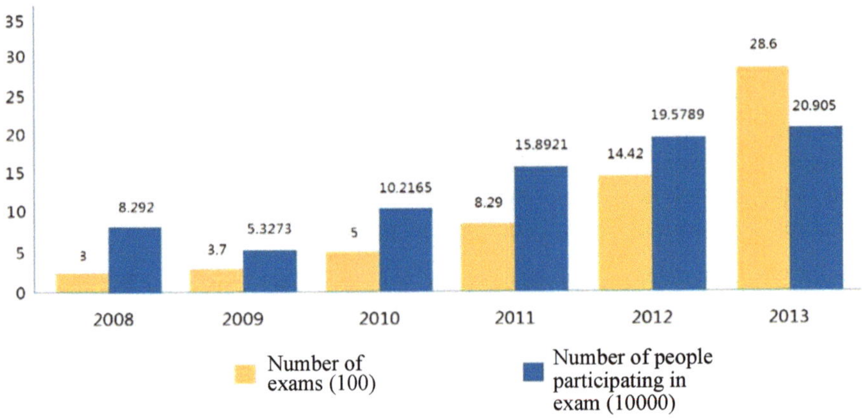

Fig. 6.5 Distribution of number of exams and number of people participating in exams during 2008–2013

(3) Develop systematic and normalized knowledge management practice activities in the whole corporate.

With the construction of 4 years, the system has constructed 6 business channels, with nearly 10,000 effective knowledge, and page view of 1,00,000 person-times every year.

(4) Provide daily simulation practical operation service relying on integrated background software.

Since the operation in 2008, relying on powerful background software and classrooms distributed in various operation points, it supports for the simulation operation for workers at the production line, to promote their operation skills and effectively reduce the error rate.

In a word, W and Z are successful examples of corporate universities, they are the prototype of future corporate universities. By analyzing the two cases, the author could tell that mature corporate universities should have perfect online learning management system, unique curriculum system and their own characteristics. Online learning management system could realize remote education as well as mobile learning, which would be helpful to popularize smart learning. Unique curriculum system should meet the learning and development requirements of the corporate, and it could be constructed based on corporate citizenship, contextual framework and core competencies (Chen 2009), which is often abbreviated as "3C". Mature corporate universities will not blindly imitate other universities because they know themselves very well and have clear direction and objective, which means they must have their own characteristics. So the future development of corporate university could use the two successful cases as a reference and discard the dross and select the essence.

6.4.4 Promotion of Corporate Online Learning Platform to Smart Learning

1. Support blended learning

The learning management system provides learners with blended face to face teaching and network teaching learning resources, which supports for not only collective learning, but also individualized learning according to personal situation.

2. Interactive learning support

When encountering a problem during learning or working, the learner can browse the knowledge base of resource management system, to find out related contents, and he can also seek for help from colleagues on the forum. In a word, the information technology function can achieve the good interaction between learners and resource management system.

3. Record learning behavior track

The learning system can record the learning course of learners, including online length, receiving and sending e-mails, public comments and online note. It is beneficial for the corporate to understand the learning behavior of learners in time, and conduct proper analysis to the learning state of learners, and give timely help in combination of the features of the learners.

6.5 Directions of Future Development of Corporate Universities

The internet is a great milestone for human historical development, and human enters into the informationized society by internet, and there have been great changes on production, life and learning. Under this background, the future corporate will transfer the traditional learning activities to the internet to breakthrough the time and space limits, to meet the requirements of learners to build up intelligent ecological system by means of informatization. The construction orientation of future corporate universities will at least include the following two aspects:

6.5.1 Platform Construction Based on Big Data Analysis

Construct big data analysis based learning platform, and by acquiring, storing, managing and analyzing of learning big data, provide digital decision plan for corporate training and education work.

Collect the interactive data between learners and learning platform, including answer correct rate, time for answers, quantity and nature for help and the repetitive rate of incorrect answer, and this part of data can be on layers of course, learning unit or knowledge point. By data mining and analyzing, it constructs learner knowledge model, and selects suitable method and provide proper learning contents at the right time for learners by automatic or manual feedback.

Collect the data of learning time, course learning completion situation, online or offline exam result and learning behavior change situation in class or during face to face learning of learners on the learning platform, and explore the correlation between learning behavior and learning outcomes of learners, to finally construct learning behavior model of the learner.

Collect the learner related data in the learning platform and the offline learner basic information data, to construct learner individual learning file by data mining and machine learning algorithm, to analyze the learning features of learners, and conduct clustering and grouping for learners with the same learning features, to provide individual learning environment for different types of learners, to promote effective learning. By the big data mining and learning analysis of education, it conducts reconstruction to the existing professional knowledge model of the corporate, to explore the correlation between course, learning unit and knowledge point learning content organizing method and learning outcomes.

6.5.2 Platform Plan Based on the Trends of Mobile Learning

The following functions shall be noticed when planning the platform based on mobile learning trend:

1. **Intelligent pushing**.

The digital learning resources in big data times will increase in an explosive way; facing with the great number of learning resources, it will be the main problem for the learners to select, or how to push suitable learning resources for learners. Based on the mining and analyzing of big data in the future, it is available to analyze the learning style, characteristic and current situation precisely, to push suitable learning resource for learners in an active way.

2. **Effective learning**.

The learning platform in the future shall be equipped with self-learning ability and the logistic interaction between knowledge, and can automatically increase the knowledge needed to solve the problem, and generate learning resources. This learning system can grow with learners, and the previous learning experience can be accumulated to answer to the question of the later learners. The solving of problems is not limited by time, space or environment, and can collect the knowledge and experience of multiple fields experts and support the mutual association between learners, to solve major problems.

3. **Immediate interaction**.

The design of future learning platform shall be equipped with interaction methods of dialog mode, interactive mode and athletic mode and so on, and can give positive response to the requirements of learners, and motivate the enthusiasm of learners. The 4G based application can ensure the immediate interaction between learners and teachers as well as between learners; it supports for individualized learning, and presents different learning resources after knowing the related characteristics of learners, and tracks learning record or process, and automatically collects and analyzes the mobile learning behavior features of learners, and establishes the individualized learning development model and gives individualized guidance based on the model.

References

Allen, M. (2002). *The corporate university handbook* (p. 9). AMACOM Div American Mgmt Assn.

Amy, L. A., & Jessica, L. (2012, March 19). Exploring the corporate university phenomenon: Development and implementation of a comprehensive survey. *Human Resource Development Quarterly, 23*(1), 103–128.

Anderson, M., & Lichtenberger, B. (2007). The corporate university landscape in Germany. *Journal of Workplace Learning, 19*(2), 109–123.

Blass, E. (2001). What's in a name? A comparative study of the traditional public university and the corporate university. *Human Resource Development International, 4*(2), 153–172. doi:10.1080/13678860121806

Chen, L. (2009). Corporate university: Background, definition and pattern. *Journal of Ningbo University: Education Science, 2009*(2), 16–21.

Chen, W., Ye, X., & Xu, Y. (2012). Future classroom: Smart learning environment. *Distance Education, 5*, 42–49.

Dealtry, R. (2000). Strategic directions in the management of the corporate university paradigm. *Journal of Workplace Learning, 12*(4). doi:10.1108/13665620010332903

Denise, R. H. (2002, May 10). *Education in the workplace: An examination of corporate university models*. Retrieved January 27, 2009, from http://www.newfoundations.com/OrgTheory/Hearn721.html

Feng, X., Wu, Y. H., & Zhu, Z. T. (2013). Design of smart learning experience. *China Educational Technology, 12*, 14–19.

Guo, X. S., Zheng, X. D., & Yang, X. M. (2014). Conceptual framework and model design of smart learning. *Modern Educational Technology, 8*, 5–12.

Ji, W. N. (2011). Corporate university on the platform of American higher education. *Renmin University of China Education Journal, 29*, 3.

Jiang, Y. L. (2012). *Research on the development of enterprise university in China* (master). Jiangxi Normal University.

Ke, Q. C. (2010). Enterprise digital learning: Concept, research framework and practice model. *Distance Education, 1*, 49–54.

Levine C. I. (1996). Corporate university or corporate college-which is right for you? *Corporate University Review, 4*(2), 52.

Levinson, M. (2007). *Knowledge management definition and solutions*. Retrieved March 7, 2007 from http://www.cio.com/article/2439279/enterprise-software/knowledge-management-definition-and-solutions.html

Li, M. L., & Wu, S. Y. (2013). Research on issues of enterprise university: Exploration and analysis of the current situation and vision of the future. *Vocational & Technical Education Forum, 33*, 36–41.

Luo, J. H. (2011). The development and inspiration of foreign corporate university. *Comparative Education Review*.

Meister, J. C. (1998a). Extending the short shelf life of knowledge. *Training and Development, 52*(6), 52–53.

Meister, J. C. (1998b). Ten steps to creating a corporate university. *Training and Development,* 52(11), 38–43.

Paton, R., Peters, G., Storey, J., & Taylor, S. (2005). *Handbook of corporate university development*. USA: Gower Publishing Ltd.

Prince, C., & Stewart, J. (2002). Corporate universities—An analytical framework. *Journal of Management Development, 21*(10). doi:10.1108/02621710210448057

Ramli, A. H. (2008). *An overview of corporate universities*. Retrieved December 24, 2008 from https://zh.scribd.com/doc/9407517/An-Overview-of-Corporate-Universities

Sham, C. (2007). *Corporate universities in China.* Retrieved February 13, 2013 from http://www.studymode.com/essays/Corporate-University-In-China-1423133.html

Tian, Y. (2013). *Research on teaching management model of enterprise university* (master). College of Education of Tianjin University.

Wu, Zh H, & Liu, X. (2013). Characteristic analysis of enterprise university and research on its mobile knowledge system. *Leadership Science, 29*, 40–42.

Zeng, X., & Liu, X. (2006). Necessity of mechanism of enterprise university institute binding. *Chongqing University (Social Sciences) No. 3.*

Chapter 7
Industry Promoting Smart Learning

Abstract This chapter discusses the developmental stage and main features of smart learning industry in China. The analyses are based on the data and facts described in industry reports published by research institutes and Internet companies (e.g. Baidu, Tencent, Alibaba, etc.). In addition, the project team of this book has carried out online questionnaire surveys to collect supplementary data. The analysis of these data shows that the industry of smart learning develops rapidly in China as a result of increasing market demand, the strengthened support from governmental policies, and the encouragement from capital investors. It cultivates three typical business models: the digitization of traditional regular education, development of learner-centered learning platforms, and creating online schools for professional training in subdivided fields. The industry chain that helps to build the flexible lifelong education system has almost reached the point of maturity.

Keywords Industry chain · Market structure · Customer segmentation · Online education · Prospect · Risk

7.1 Introduction

During the last decade (2000–2010) smart learning in China was more of a concept. Smart learning started its industrialization process since around 2010, motivated by both government policies and investment capitals. Online education has greatly impacted the traditional education market (Zhiyan Consulting 2013). With the popularity of the Internet and mobile devices, people are used to search for and obtain information through the web, and there is a rising demand for continuing education and skill training among working people. Companies and investors taking smart learning as a promising industry have used the opportunity to penetrate and occupy the market and the whole industry chain has been formed. Platform-based enterprises play the leading role in the process of industrialization, reshaping the business ecosystem, and squeezed the living space for smaller

D. Liu et al., *Smart Learning in Smart Cities*, Lecture Notes in Educational Technology, DOI 10.1007/978-981-10-4343-7_7

companies. A trend of overheating has appeared as capital flooded into online education market around 2014, statistic data shows that over 60% of the projects invested by capital since 2013 has been shut down by 2015. Above facts show the immaturity of the market and the blindness of the capital, thus besides the self-improvement of the market, the government should enhance the guidance for capital investment in this area.

In order to better understand smart learning in smart cities, the industries that support smart learning should be analyzed. The industries and capitals that support smart learning play an important role in creating of the ecosystem for smart learning in smart cities. The following sections will introduce the development status of smart learning industries, the maps of capital investments for smart learning, and the future trends of smart learning industries.

7.2 Developmental Status of Smart Learning Industry

The development of smart learning industry can be analyzed from three perspectives. Firstly, as the support of high-end services of smart city, as it provides learning tools that are terminal-based and intelligent. Secondly, the industry drives the overall upgrading of learning behavior. The capacity of sustainable development and comprehensive competitiveness of industrial enterprises contribute to industrial optimization. Thirdly, the key to industrial breakthrough is in forming of the smart learning environment in the whole society.

The content of this section is based on the results obtained by the "Questionnaire Inquiry About Online Educational Industry Conditions" (QIAOEIC) utilizing so called "sojump" (an online survey tool widely used in China) and carried out by the project team in May, 2015.

7.2.1 Full Coverage of the Lifelong Learning System

"Establishing flexible and open lifelong education system" is the direct manifestation of smart city development. Such system breaks the boundaries between degree education, non-degree education and community education, and gets through the fence between schools and learners. It makes use of educational resources more enjoyable for learners, and inspires the enterprises and institutions specialized in educational industry to create learning resources covering the whole life cycle.

(1) **The smart learning industry is mainly involved in 7 areas**: preschool education, K-12 education, college and university, vocational school, enterprise, government, and personal ability enhancement.

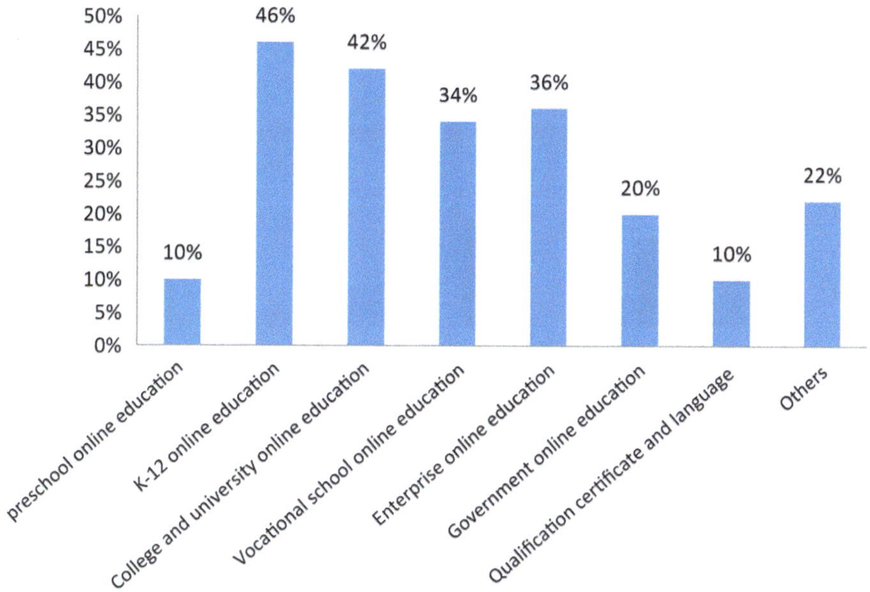

Fig. 7.1 Proportion of smart learning industrial enterprises and institutions in range of services

Among these 7 areas, the K-12 education takes the majority; meanwhile, the online education for the government and for employees of corporate business start to get attention of industries (see Fig. 7.1.[1])

(2) **The traditional education market is impacted by online education**. Each segment of education is impacted by online education to different extents, with higher education and vocational education are impacted deeper. Since the Internet, the PCs and intelligent devices have been playing more important role in people's daily life, they are more open to online education. This trend can be seen in the compound annual growth rate of market scale of online education from 2013 to 2015. (Online Education Market Size in China and Year-on-year Growth Rate in 2013–2015, 2013), as shown in Fig. 7.2.
(3) **All the 7 areas incubate clear market path**. "Preschool education" products occupy an important position in APP store education section. The "K-12 online education" relies on the high-quality school resources and online education brand. "MOOC" model of higher education is further practiced to make higher education more equal. Vocational education has a good development tendency, and it is a new market of education. Some enterprises and institutions with the annual income of more than CNY 100,000,000 has emerged in Enterprises'

[1]Data source: White Paper Project Team of Smart Learning Institute of Beijing Normal University, samples: 50 > middle-high administrative staff from online education industry.

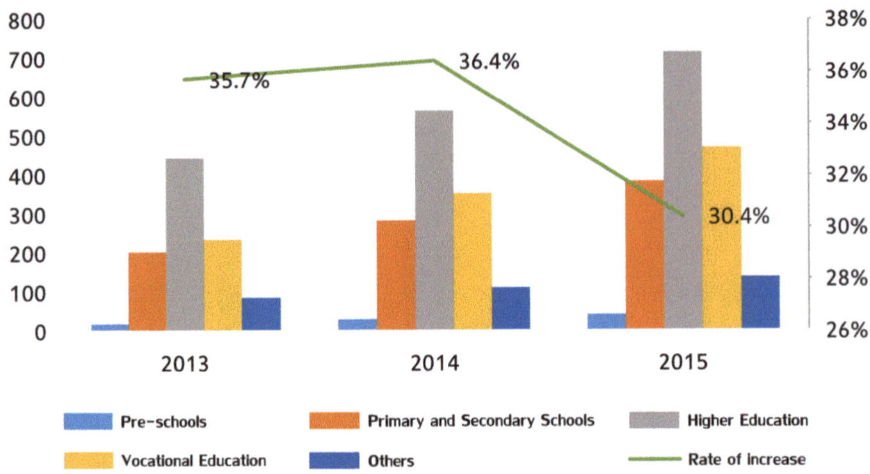

Fig. 7.2 Online education market size in China and year-on-year growth rate in 2013–2015

online education. Government's online education becomes the new emerging form in online educational areas. The enhancement of personal ability by online education covers older learners and social learners, etc.

7.2.2 Increased Diversification of Industry Solutions

Currently, with the platform-based enterprises such as BAT (Baidu, Alibaba and Tencent) invested in smart learning, the types of industry solutions are increasingly diversified in China with the continuous capital inflow and the improvement of user penetration and acceptance.

(1) **The smart learning industry enriches the connotation of smart city industry**. Smart learning industry is the foundation for the sustainable development of smart city. The strategic emerging industry is with intensive knowledge technology, less resource consumption and large growth potential. It contributes to constituting modern industrial system and promotes the overall economic and social development. In the QIAOEIC carried out by the project team, the enterprises and institutions serving the learning environment and supporting the services and learning resources take up a large proportion in all the industrial solutions for smart learning (see Fig. 7.3).

The industrial solution mainly includes the following five types:

- Smart learning environment that is composed of broadband network, smart campus and smart classroom;
- Learning facilities that include the intelligent terminals such as the terminal of mobile learning, mobile storage device and wireless network equipment, etc.;

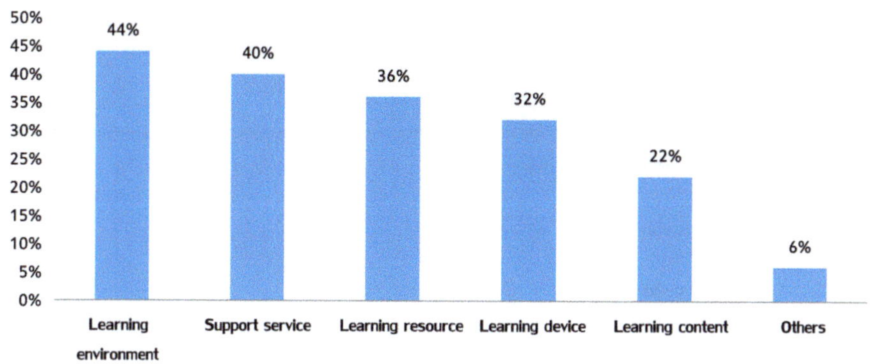

Fig. 7.3 Types of smart learning industrial solutions

- Learning resources that include community education resources, television program, open university, etc.;
- Learning content of video production and the learning content platform;
- Supporting services that include platform-based portal supplier, smart learning consultative agency, etc.

With the aggravated industrial segmentation and expansion of vertical areas, the types of industrial solution will continue to expand.

(2) **The online learning products need further development**. Currently, the research on online education concentrate mainly on 3 aspects: business model and technology, user experience and measurement of learning effects, control and description of learning products, personnel organization, and learning process. It was found that the product forms mainly include B2B2C platform-based form, B2C service-oriented type, tool-oriented type and online class-oriented form (Wang 2014) (see Table 7.1).

Table 7.1 Main product forms of online learning

Product form	Characteristic	Example of typical enterprise and institution
B2B2C platform-based form	Cooperate with institutions. Personal teachers enter. Provide online and network teaching resources	Chuanke.com, Dauber, 51CTO
B2C service-oriented	Produce high-quality contents independently	Kuxuexi.com
Tutorial tool-oriented	Answer the test paper. Design the test paper intelligently. Do exercises by passing through a series of difficult levels	Yuantiku.com, 17zuoye.com
Online school-oriented	Real 1 to 1 teaching-student video guidance	91waijiao.com, New Oriental

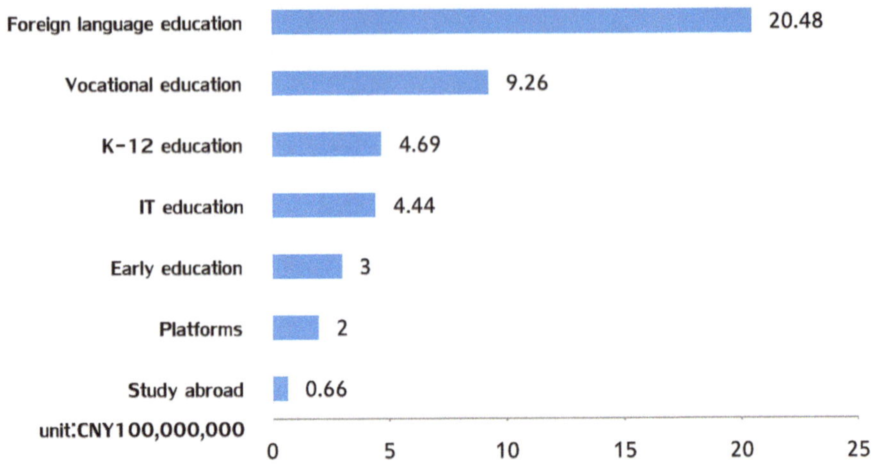

Fig. 7.4 Investment amount in 7 educational enterprises in 2014 (*unit* CNY100,000,000)

(3) **The capital market is more optimistic about vertical online educational products**. The platform layout ability of vertical online educational products is inferior to that of the Internet enterprises. However, the vertical enterprises generally possess core competitiveness: teacher resources, mature teaching charge profit model and user habits. According to the *K12 Educational Market Analysis Report in 2014* published by Tencent Class, the in online education exceeds CNY 4.4 billion, including CNY 2 billion in foreign language, CNY 0.469 billion in K-12 education. Despite the outburst of K12 in 2015, this market is still facing severe challenges such as low maturity of user, competition homogeneity, and non-standardization of service, etc. (see Fig. 7.4).

7.2.3 Promotion of Learning Society by Intelligent Tools

On August 17, 2013 the State Council of China issued the "Broadband China" strategic implementation scheme, which deployed the development goals and path of broadband for the next 8 years. It means the "Broadband Strategy" has been upgraded from a sector's action to a national strategy. Broadband becomes the national strategic public infrastructure for the first time. The "Broadband China" strategy and network infrastructure are deepened, which provide space for the development of various smart devices. From smart phone to Tablet PC and wearable devices, the intelligence and networking of terminal products creates the explosive opportunity of smart learning industry in the new economic environment.

(1) **Learning products should provide a good human-computer interactive experience**. The mobile smart terminal products design needs the full consideration of user experience. The convenience degree of learning products directly influences the experience effect of learners. The convenience of learning product is mainly manifested as to whether the human-computer interface is friendly, whether the operating product is easy to use, and whether the relevant help and support is available. According to the statistics resulting from the QIAOEIC, the very high and high satisfaction for the convenience of learning products of the investigated institution take up 68% (see Fig. 7.5).
(2) **There are three universal evaluation standards for excellent learning products allowing human-computer interaction**. The first standard is priority that learning products should display the most valuable resources to promote user-interface interaction. The second standard is consistency. The interface is easy to be predicted. It shall reduce the cognitive and learning burden of users. The third one is experience. It makes the learners experience the convenience and safety of learning through high-quality learning products.
(3) **The learning products should support personalized learning**. Personalization is the main feature of the future lifelong learning. The personalized learning needs learning products to be adaptable to various differences in demand, and needs to provide different customized learning services for learners. If the personalization degree of the product is higher, learners feel to be more valued and develop stronger learning motivation for learning (see Fig. 7.6).

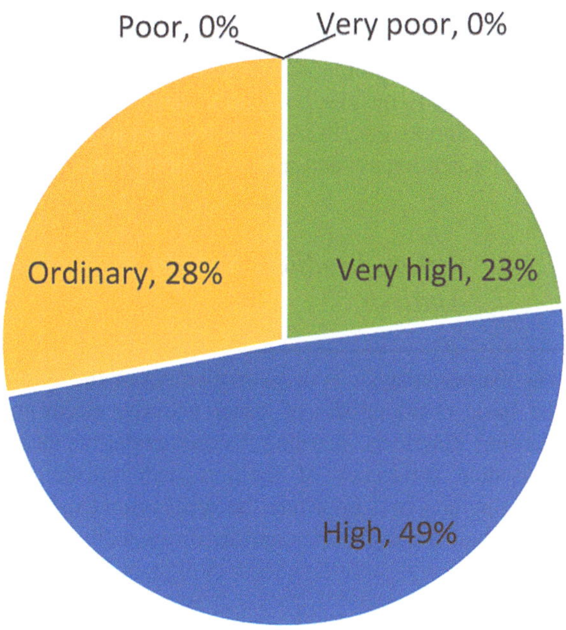

Fig. 7.5 Satisfaction proportion for the convenience of learning products

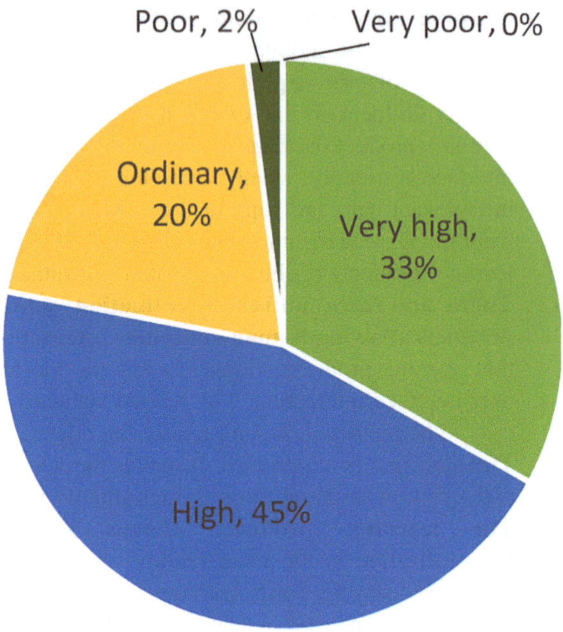

Fig. 7.6 Recognition degree of personalized design of learning product

(4) **The online learning products need sophisticated design**. It is very clear for the learning content, learning objective, learning plan in traditional classroom offline learning, so no complex theory or design skills is needed for such a learning. However, online education needs the comprehensive consideration of various links of learning due to lack of enforcement and discipline, so as to guide the participating enthusiasm of learners. The design of learning products shall also consider cognition, memory, imitation, exercise, understanding and application, etc., to motivate learners to generate learning behavior actively and conscientiously, and achieve learning purpose.

7.2.4 *Online Products Should Be Optimized Through User Analysis*

The learner analysis is the core element of the smart learning environment. One of the basic characteristic of smart learning environment is providing personalized learning diagnosis, learning recommendation and learning service based on the individual difference of learners and individual records of learning process. Learning analytics improves the application experience of product via data analysis, and solves the difficulties and doubt as well as problems during learning.

(1) **The product-replacement rate is manifested in the research and development ability of online learning enterprises**. It can reflect the research and development ability of innovative enterprises. If the product upgrade speed of an enterprise is high, it will make a favorable impression of on customers, and gain their affirmation and support %. Generally speaking, the upgrade speed is high (see Fig. 7.7). However, according to the research data the products updated within half a year take up 54%, and those updated within 1 year take up 26.
(2) **Product upgrade promotes industry upgrade**. In the last three years, the online education field has been surging forward vigorously, but cannot avoid problems such as conceptual speculation, misinterpretation of online learning and disdain of learning product design. The upgrade of online learning products could promote the improvement of the eco system of learning industry. For example, one online product remodels the traditional video into the "flipped class" model for learners' self-regulated learning by inspecting the learning effect, taking part in live interactive exercises for improvement, and answering questions and providing solutions at any time for 7 × 24 h.
(3) **Learner analysis is the core factor for the growth of smart learning industry**. The trend of online education is to provide personalized and customized learning services for students according to standard algorithm, system model, date mining and knowledge repository. Based on such services, the specific learning behavior and teaching behavior can be analyzed what can improve the teaching techniques used by teachers, and upgrade the learning behavior of learners. According to the data obtained by the QIAOEIC, the

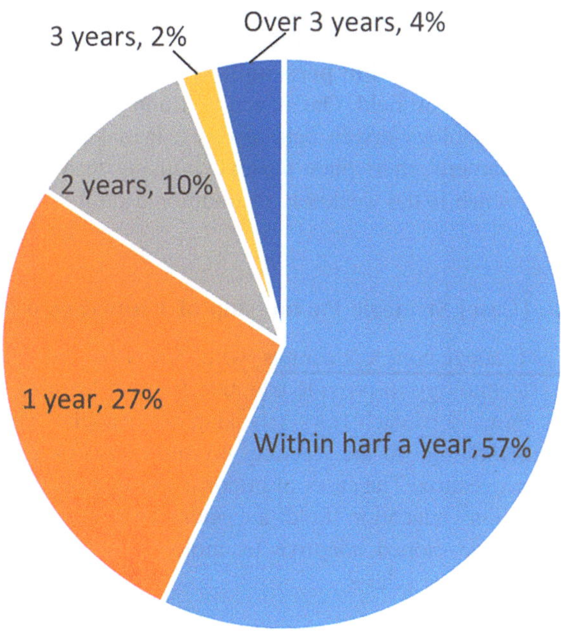

Fig. 7.7 Upgrade speed of learning product

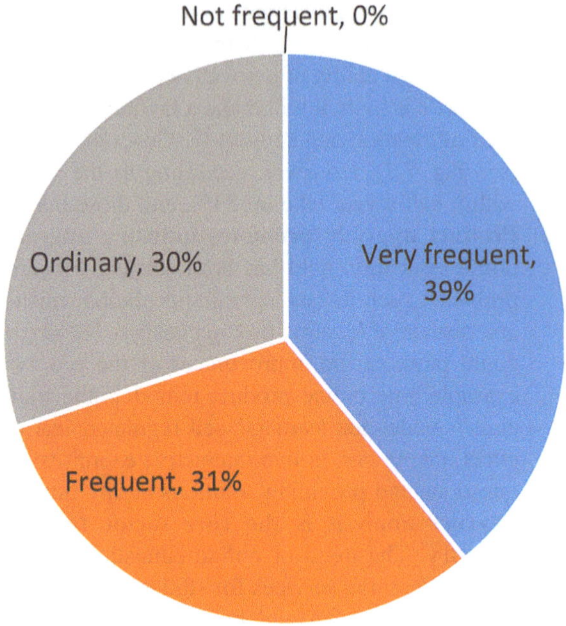

Fig. 7.8 Frequency of learner analysis

proportion of enterprises and institutions that very frequently or frequently conduct learner analysis reaches 70% (see Fig. 7.8). It indicates that in the era of information technology, understanding the user's demands and improving product service through the data mining have become the consensus of most enterprises or institutions.

(4) **Technology for personalized data analysis is in great demand in online education field**. One new opportunity in online education market is the field of technology-driven data analysis. It makes instruction more efficient aiming at learners' short plate through infusing technology into instructional guidance, which is the weakness of traditional offline education and one-to-one tutoring.

[Case] Strategic Partnership of 17zuoye.com and Knewton

17zuoye.com cooperates with Knewton that possesses adaptive learning technology to provide adaptive English learning in China. The personalized demand of each learner is adapted. It applies the personalized data analysis technology to the teaching products through data collection, deduction and suggestion. The cases of importing personalized data analysis technology into online education field are increasing. "100 Education" also imports its self-developed adaptive technology into the vocational education field of global net-school.

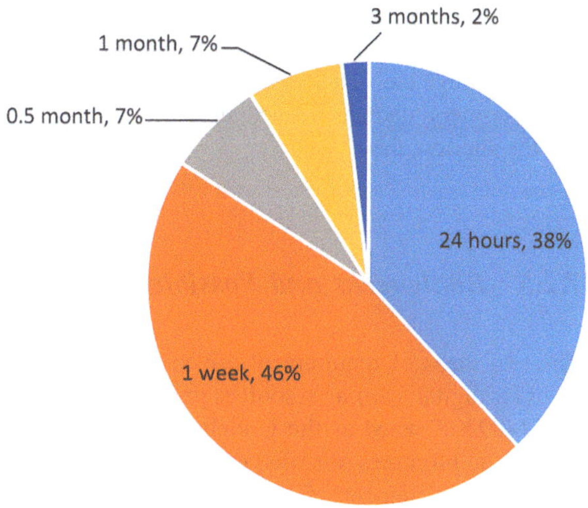

Fig. 7.9 Speed of online education enterprises and institutions to respond to learners' questions

(5) **The speed of solving problems for online education enterprises is becoming important**. Low response speed will reduce the customers satisfaction and decrease their intention to use the service again. The studies on enterprises and institutions offering online education, show that 38% of online education enterprises can respond to learners' questions within 24 h and 46% of them require one week for providing a solution. It indicates that the speed and efficiency of enterprises of learning products to solve learners' questions need to be improved (see Fig. 7.9).

[Case] "Wisdom Teaching Assistant" Robot of Huayu Education

The robot "wisdom teaching assistant" replaces teachers in the classroom, reads textbook content according to curriculum, and solves students' problems at any time. Through its built-in "elite-teacher class" application, students can make remote English learning interaction with foreign teachers in the way of video call through the tablet PC in the robot's head. It solves the problem of "difficulties of innovation teaching, interactive learning and parents' involvement. "Smart learning class" will become the principal model of K-12 classroom teaching in the next 5–10 years.

(6) **The learning system should rapidly respond to the questions and demands of customers**. Online education results in the change of learning behavior: the flipped class gives the initiative to students; answer questions through mobile phone photographing; machines start to infiltrate into the interactive dialogue

summary; hardware is used for educational training, such as understanding the physical world through robot programming; the interaction improves learning efficiency, etc. The learning system relying on technology can respond to customers' demand immediately and provide the required learning resources in the shortest time.

7.2.5 Bottlenecks and Possible Solutions

The two ideas of requirement-driven application and application-driven development for digital learning resources proposed in "*National Plan for ICT in Education (2011–2020)*" acted as the guidelines for smart learning market. The Ministry of Education promotes the fusion between information technology and education through "requirement-driven application". It issued a series of documents that provide guidelines for the establishment of online open courses and public service platforms. These applications become the new growth point of the industry.

(1) **The online learning of K-12 Education should apply O2O**. There are several important dimensions to judge the core competitiveness of online education, including concise application, high-quality content, smooth channel and diversified user experience. No matter whether it is K-12, vocational training or studying language before going abroad, more and more students adopt the learning model that includes the Internet products. However, the younger is the age of learners, the lower is their ability for self-regulated learning. Young learners often cannot complete the online courses independently, which directly influences the judgment to online education effect. According to the conclusions of "penguin chi cool" report (Internet cross-border report of second and third class cities in China 2015a) and data from the research of our project team, it can be seen that although online education is flourishing, the satisfaction of users is rather low (see Figs. 7.10 and 7.11).
(2) **Scientific assessment of learning effect contributes should break the bottleneck of industrial development**. There are four levels for the evaluation of learning. The first one is evaluation of reaction, the second one is evaluation of results of learning, the third one is evaluation of behavior, and the fourth one is evaluation of performance. The statistical data obtained by the QIAOEIC way indicate that the key to overcoming bottleneck of industry development is in evaluation of learning effects of online learning courses (see Fig. 7.12).
(3) **Application effect weakens the industry innovations**. In contrast with the flourish of the industry, the popularity at the user level is relatively lag behind. On the one hand online learning industry is still trapped in the lack of interactivity of its products that can not trigger learner's motivation and enthusiasm. On the other hand, the industry is still in the initial stage, and the application level requires further popularity. As for the evolution status of comprehensive

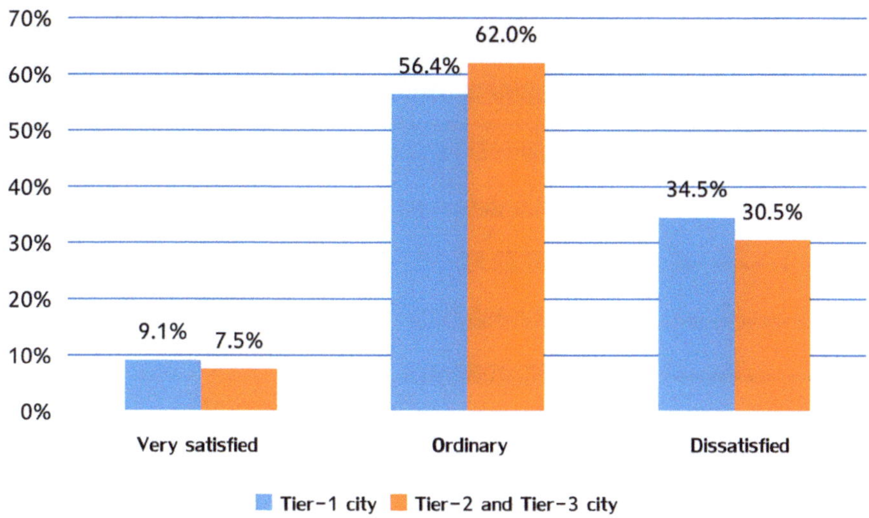

Fig. 7.10 Satisfaction of users in different cities to online education in 2015

Fig. 7.11 Customer satisfaction self-assessment for online learning products

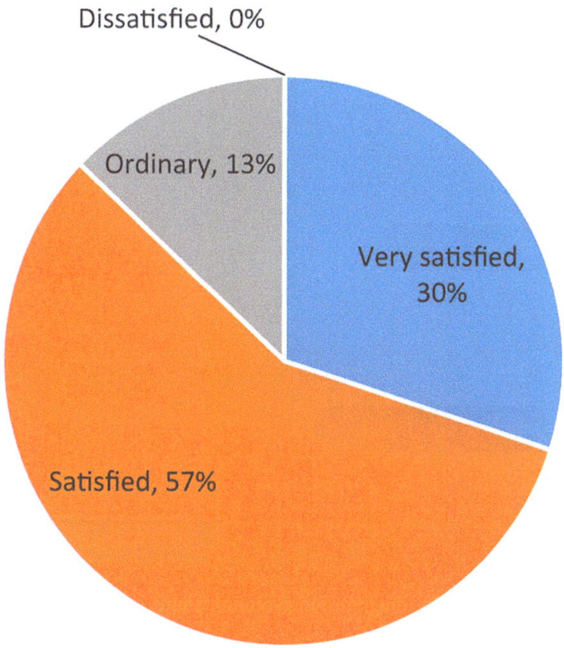

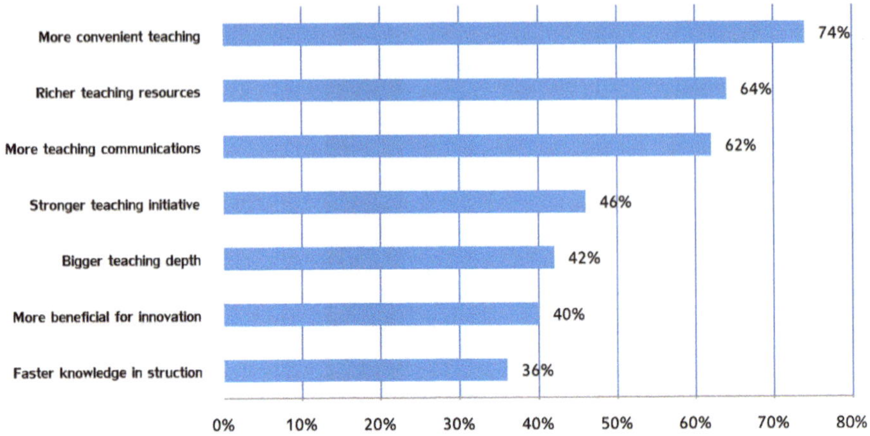

Fig. 7.12 Application effect of online learning product

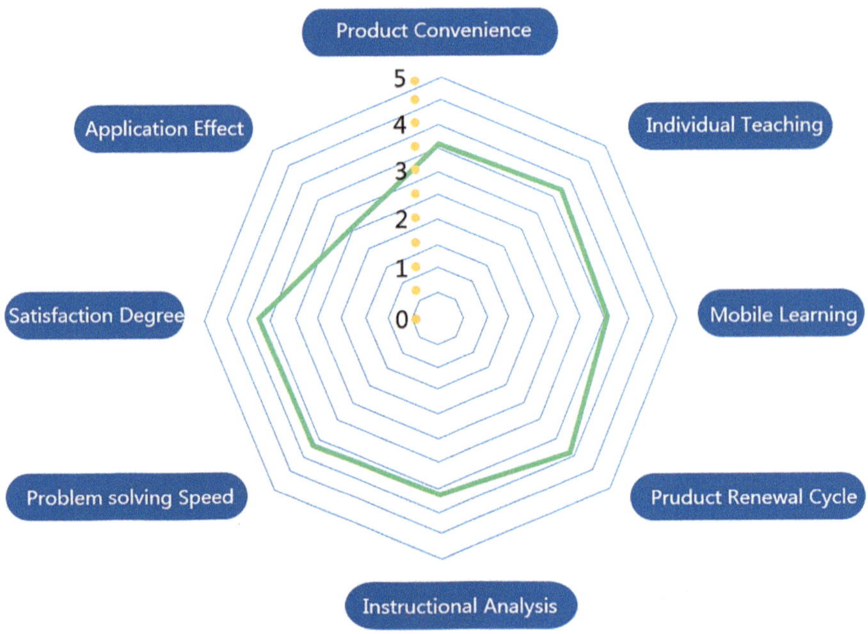

Fig. 7.13 Radar diagram for evolution status of smart learning industry

smart learning industry, the application effect is the weakness of the whole industry. It is a huge challenge to know how to utilize online learning products and improve the expected effect of learners[2] (see Fig. 7.13).

7.3 Map of Capital Investments Supporting the Smart Learning Industrialization

Capitals (investments) have penetrated into every aspect of Chinese education. The statistical data indicate that in the last 10 years there has been a rapid expansion of private education. The educational digitization industry grows at a compound annual rate of 20%. With the popularity of the concept of information society and learning society, capital cooperates with smart learning industry more closely. Therefore, the industrial capital map is formed rapidly.

7.3.1 Open Capitals Getting Involved

One question frequently discussed in the domestic educational circle is what kind of knowledge students need to learn in the Internet era. In the 21st century, the global village constituted by the Internet amplifies the value of learning. Learning has become the foundation for understanding the modern world and adaptation to the society. Learners are moving into the new learning environment, where the industry organizations invested capital to different extents for profits. Smart learning enterprises constantly optimize the learning environment to obtain more market value and business share.

(1) **The industrial chain of online education has been formed** The industrial chain of online education mainly consists of platform suppliers, technology suppliers and content suppliers. One forecast report issued in 2014 by IBIS indicated that the platform, technology and content suppliers account for market shares of 70, 21 and 9% respectively. Though platform gains the upper hand at present, the content will probably play a dominant role in accordance with the future trend (see Fig. 7.14).

The overall netizen scale continues to expand. Online education habit of users is initially formed. The Internet environment is gradually popularized and the mobile Internet is well established. The emerging Internet market with hundreds of billions output has been rendered. Users' demands will directly stimulate the market. Chinese online education market will remain at over 30% growth rate per year in

[2]Data source: White Paper Project Team, comprehensive statistical scores in "Questionnaire Inquiry of Online Education Industry", May 2015.

Fig. 7.14 Online education industrial chain

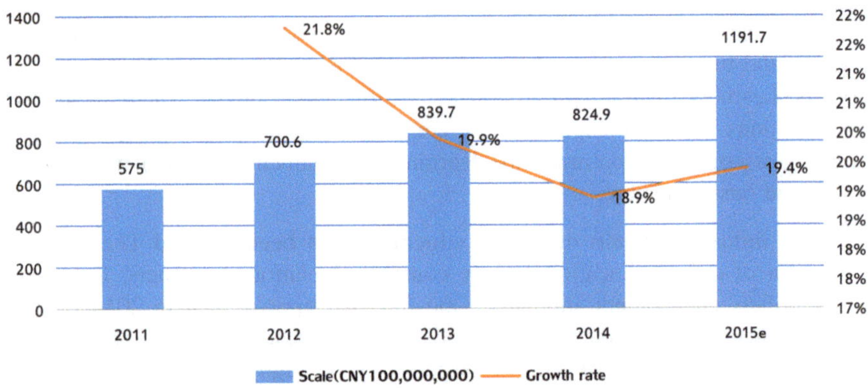

Fig. 7.15 Market scale of Chinese online education

the next few years. The growth rate ranks the second in the world. The market scale is expected to exceed CNY 170 billion in 2017 (see Fig. 7.15).

(2) **Online education has become the new economic growth point**. The financial input into online education in 2014 exceeded CNY 4.4 billion, which resulted

in some obvious changes: the requirement of degree programs decreased, language training courses were getting increasing attention, and the awareness for K-12 education rose steadily. Online education products of language training and K-12 education developed rapidly.

(3) **Online education gains the favor of the capital investment**. The capital's awareness is a key indicator of a certain product or an industry's value. Investment in online education in 2012 presented a trend of compound growth. Online education platforms have become the hot spot of capital market, such as SmartStudy invested by Baidu, VIPABC invested by Alibaba, "Wanmen University" invested by Renren company, and the 91waijiao.com invested by Netease, etc. The intervention of capital market has brought new vitality for this industry. Some investment cases in 2013 and 2014 indicate the attentions of capital to this field (see Fig. 7.16).

(4) **The cross-border trend of smart learning industry is obvious**. BAT (i.e. Baidu, Alibaba, Tencent), as the big three market players firstly seized the business opportunity and entered the field of smart learning, BAT adopted the "self-establishing + investment" way, bringing funds, talents, technology and ideas for the industry. Smart learning industry is a developing sunrise industry with bottleneck and deficiency, which needs the common participation of social forces for solutions. Cross-border enterprises bring new ideas and thoughts, promote resources to flow naturally towards the value low-lying land of learning industry. Four main types are formed: platform-based trans-boundary, central trans-boundary, value-added trans-boundary and integrated trans-boundary (Fig. 7.17).

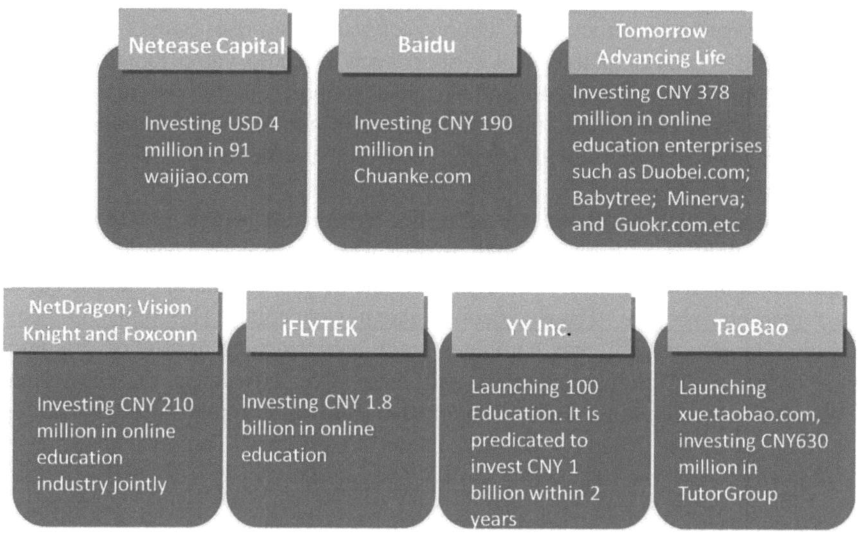

Fig. 7.16 Several investment cases of internet enterprises and education institutions in 2013–2014

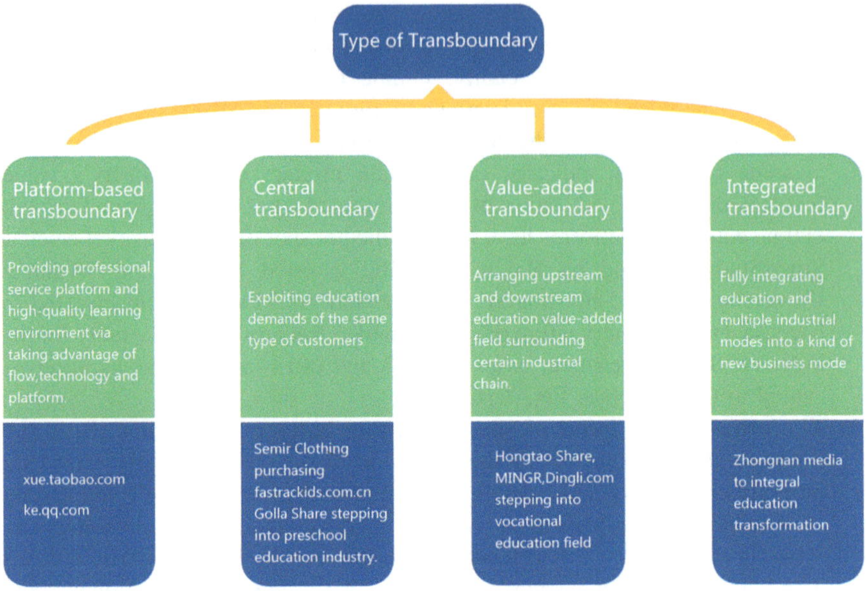

Fig. 7.17 Main cross-border types of online education industry

(5) **The business type of A share market presents 5 echelons**. From the perspective of main enterprises of online education concept stock, enterprises involved in online education in the capital market mainly include: enterprises specialized in traditional home-school interaction field, enterprises participating in information construction of schools, software enterprises based on education, enterprises providing technical support such as audio video tools, and enterprises providing broadband and terminal support. As for the representative enterprises, there are traditional enterprises in non-educational field, and also the Internet enterprises with leading technology, with an obvious cross-border trend (see Table 7.2).

(6) **The diversified cross-border of online education represents two models**. One is light cross-border model. It sets about from users' demands, without store input in short term. It is favored by the investment market and the typical representatives are 91waijiao.com, jikexueyuan.com.; The other is heavy cross-border model, that combines traditional teaching resources, methods, theories with Internet channels and innovative experience. It reconstitutes teaching and learning model through 020 innovation modes. The typical representatives are Xueda Education, New Oriental and Tomorrow Advancing Life.

Table 7.2 Five types of online education concept stock in a share market

Type	Content	Representative Share
Type 1	Traditional enterprises specialized in home-school interaction field possess the channel advantage for in-depth contact with schools and parents. They are expected to cooperate with the content enterprises in the future	Qtone Education, Talkweb
Type 2	Participate in the information construction of the Ministry of Education and schools, and provide the cloud-based platform and relevant technical support	Tianyu Information, Huaping Stock, Central South Media
Type 3	Develop more online education contents based on educational software and positive transformation	Fangzhi Technology, New Southeast Asia, Gehua CATV Network, Sumavision
Type 4	Provide technical support for audio video tools	InterPhonic
Type 5	Provide broadband and terminal support, and establish smart education platform	Dr. Peng

7.3.2 *Industrial Shuffle Motivated by the Capital Participation*

With the acceleration of the digitization in education industry, the concept of "Internet+" gradually enjoys popular support. Online education attracts capital investments. Meanwhile, the high dispersion status of this market provides good opportunity for capital to cultivate the biggest online education companies. Capital seeks the "quasi-oligarch" that may get the monopoly in certain field. The industrial shuffle is aggravated.

(1) **Online education industry enters the new stage with the coexistence of opportunity and crisis**. When the capital is flooded in, the crisis of smart learning industry starts to emerge. More and more enterprises and investors outside the education field enter the market, thus resulting in waves of failures. There are over 30 failure cases in online education filed in 2014. When users are not fully cultivated, capital inflow promotes enterprise and institutions to follow, and the Matthew Effect of the industry gradually emerges. Enterprises with potential gain multi-round investment, while institutions without profit model or capital background are in difficulties.

Since 2013, 692 projects have constituted the panorama of online education industry in China. 89 online education projects have gained investment with the investment amount over USD 560 million and 57 of these projects have stopped operation (see Fig. 7.18). The Internet severely impacts the offline training organizations. In the context of market, it is predicted that the scale will exceed 10,000

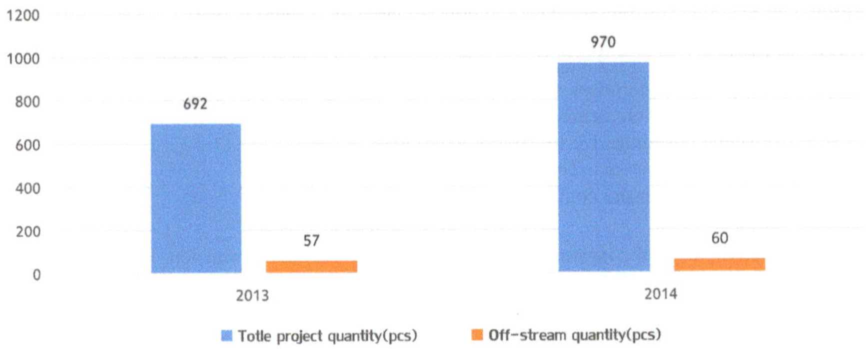

Fig. 7.18 Contrast between total project quantity and off-stream quantity of online education in 2013–2014

in 2015. How the capital injection can improve the education industry will be an ever-existing topic.

(2) **Capital oligarchy effect is initially manifested**. The development of online education is in a subtle transformation period in the information society. The intelligent hardware, Internet of Things (IoT) and big data, etc. have not been truly popularized and applied. However the mobile lifestyle of O2O and APP, etc. and innovation model are becoming things of the past. The smart learning environment taking online education as the main manifestation is one the fields that are not fixed by patterns. With the aggressive cross-border of BAT, the enterprises with educational gene will compete with the enterprises with Internet gene at the levels of operation mode and profit mode, etc. Under the impact of the Internet, industries are divided and demoralized gradually. Capital begins to put its efforts into oligarch cultivation in this field.

[Case] VIPABC where Alibaba as the leading investor gaining the largest financing case of online education

Alibaba, Temasek and Qiming Venture invested USD 100 billion in the Round B financing of VIPABC in the beginning of 2014. VIPABC provides a teaching environment that takes good students as the center. It has provided over 5 million online courses with live teachers and high-quality nature language teaching environment for more than 40 countries. Meanwhile, VIPABC can solve the issue of teachers, narrow the gap between urban and rural education, and satisfy rural learners to obtain good education and provide them with the opportunity to understand the world in a more convenient way.

Alibaba ranks education industry in the first place among the list in the investment plan in the next decade; Tencent is engaged in creating a mature remote education system; Baidu will complete market segmentation by the capital injection in Chuanke.com. As for the huge market and particularity based on online learning industry, the model dominated by only one strongest oligarch still remains to be seen. However, it cannot be denied that the intervention of capital market will accelerate the maturity of online education market. In addition, the closed-loop ecological construction-oriented enterprises and institutions will appear on the E-business platform that holds the advantages on flow and entrance.

(3) **Industrial shuffle will continue**. When online learning industry expands rapidly, the homogeneous competition is increasingly fierce. Every day 9 online education projects and 2.6 new APPs are released. In the "only the first, no the second" mobile Internet age, the competition in online education field is more cruel than in other fields. In the *Research Report on Chinese Education APP Industry Development and Users' Behavior in 2014* released by Sina Education, the educational APPs for foreign languages and exams are most frequently used (see Fig. 7.19). By the end of 2014, there were more than 70,000 educational APPs, which rank the second among the application types, after game application in the mobile phone application stores.[3]

(4) **The deep perpendicularity accelerates the industry differentiation**. Internet has brought high homogenization of products and services. The homogenization phenomenon is especially obvious in online learning field. Meanwhile, competition forces enterprises and institutions to make constant transformation, subdivide users' demands, and select the direction with large quantities of user groups for intensive efforts. In such process, industrial chain accelerates the differentiation. The division of labor is more and more elaborated in such a way that it forms durable and irreplaceable core competitiveness.

(5) **Industry shall return to the ultimate source of education**. The important mission of the online education is to motivate people for learning, exploration of knowledge and pursuit for happiness. However, the educational behavior derived from the Internet at present puts too much emphasis on teaching as about 99% of enterprises still stay at the teaching level. The learning behavior is highly utilitarian. Online learning shall change the traditional learning concept and learning behavior via advancing the quality-oriented education, to make learning the basic skill of each person.

7.3.3 Capital Is Interested in Profit Model of Online Education

The key issue for online education at present is the lack of clear business and profit models. The profit model is relatively complex. As for the future industry structure,

[3]Data source: Sina, China's Education on Fingertips—Education APP (Application Software) Assessment Report, November 2014.

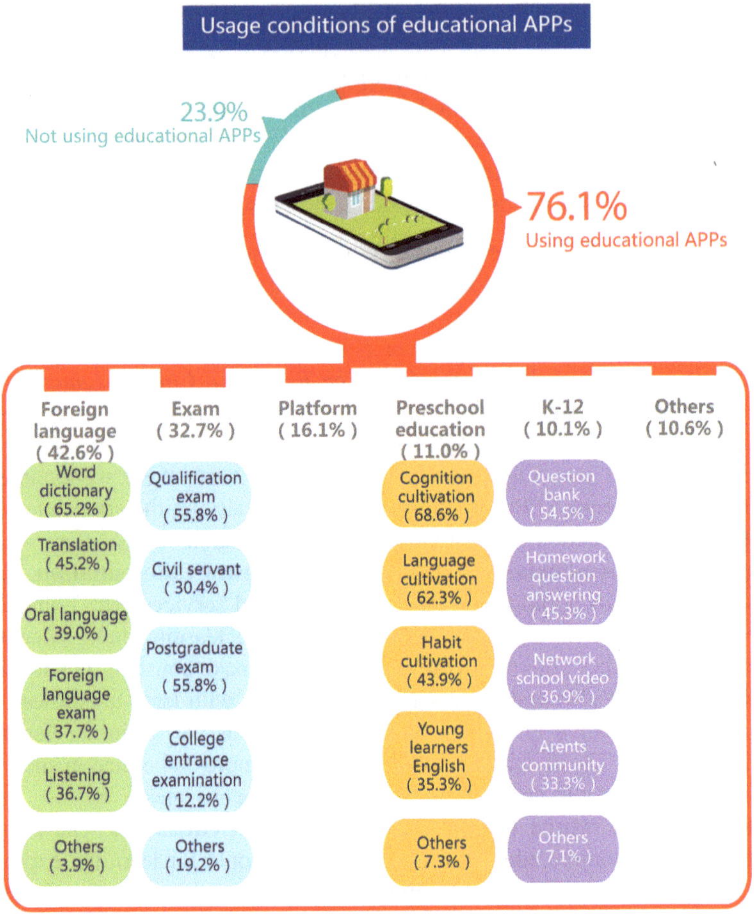

Fig. 7.19 Usage conditions of educational APPs by classification

diversified situation will appear, presenting "platform + vertical industry" dual-wheel-driving state.

(1) **New profit models are created on the basis of traditional charging model**. The profit of domestic online learning industry mainly relies on five charging models: content charges, service charges, software charges, platform commissions and advertisements[4] (see Fig. 7.20). To attract users, learning products launched by many platforms are free of charge. But, free-of-charge has no

[4]Data source: Industry Information Network, Research Report on Network Education Market Investigation and Investment Potential in China in 2012–2013, 2016.

practical significance to the sustainable development of enterprises. With the industry popularity and the formation of users' habit, endow products with payable value based on the original platform to provide quality service will become the profit points.

(2) **The differentiation of profit models originates from the variation of business models**. Mobile terminals are superior to PC terminals in terms of learning progress, voice interaction, interface design, etc. Users possess the awareness to pay for high-quality contents. The main profit points of platform-based enterprises and institutions are proportional to the distribution of income and advertisement. The charges for educational products mainly depend on course contents. Each industrial subject is exploring new profit path on the basis of traditional charging model.

(3) **The main characteristic of industry is long term for profit**. Though it is actively welcomed by capital, the overall profit model of online education is still under exploration. Many new startups generally adopt the path that firstly gains customers and then considers profits. The high cost of obtaining customers has become the core reason. The cost for obtaining customers in online learning generally takes up approximately 30% of total cost, nearly the same with the cost of obtaining customer of the offline traditional learning institutions.

[Case] MOOCS: positively exploring new profit model

MOOC stands for massive open online course, which reflects the core of online education to certain extent: to solve problems via knowledge, and gain cashability via high-quality service. MOOC platform enterprises are still exploring more feasible profit model: Lynda adopts user subscription model; Courser gains profits via course-completion certificate. All its online courses are free of charge. Certain cost shall be paid if the course-completion certificate is needed after all the courses are learned. In addition, the payable counseling on job-hunting, degree awarding and advertisement, etc. are the profit models positively tried by MOOC.

(4) **K12 profit model includes two levels of teaching and learning**. From our analysis via Kirkpatrick Model and product perpendicularity, the products in the current market can be divided into 4 types of point-cuts: teaching assignment, exercise, and tutorial (Power Education 2014) (see Fig. 7.21).

(5) **The Chinese online education industry is still at the stage exploration for the product model**. Product model determines the profit model. The product model of "teaching point-cut" mainly serves teachers. Teachers represent the teaching faculties in school organizations, which mainly manifest B2B (Business to Business) business; "assignment point-cut" serves both teachers and students, which manifests B2B and C2C (Consumer to Consumer) business; "exercise point-cut" mainly serves families and students, which manifests

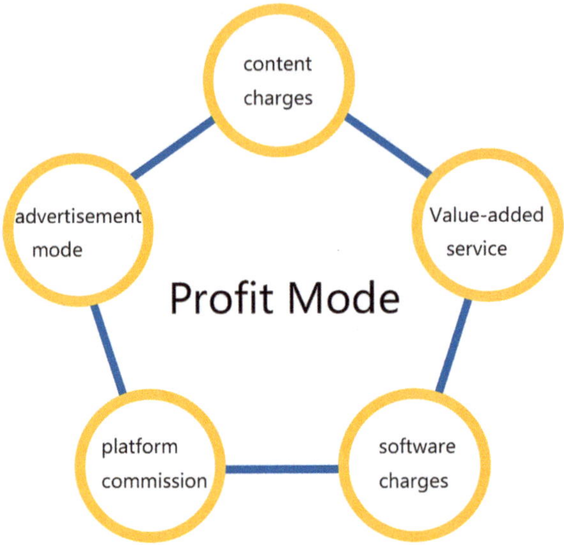

Fig. 7.20 Types of profit models of online learning industry

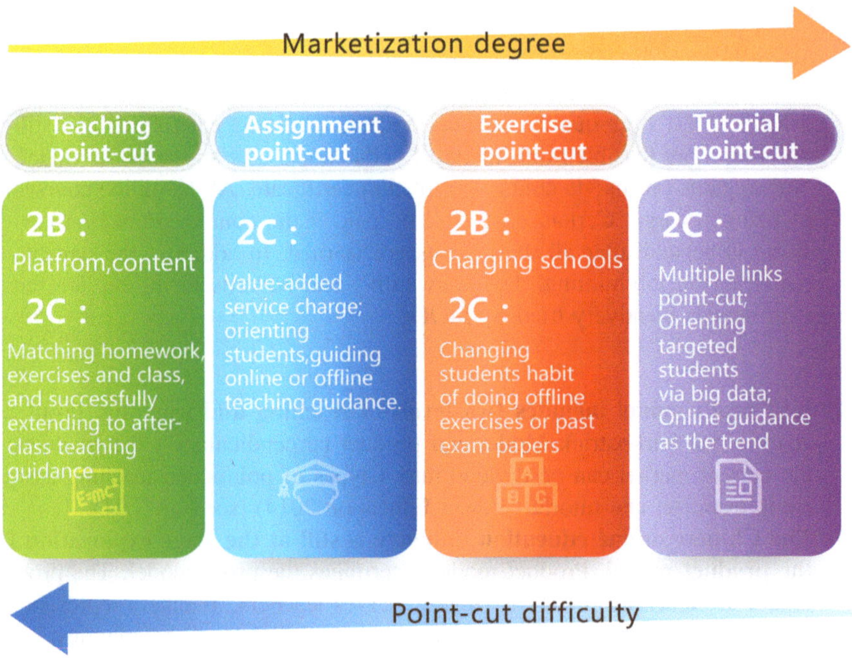

Fig. 7.21 Profit model of point-cut products

C2C business; "tutorial point-cut" gets through online and offline. Online is to position learns' behavior and purpose, and offline is for orientation guidance, assessment and inspection. The industrial chain of the product form that takes school teaching as the point-cut is the longest. No matter which kind of model, there will be steady profit model only after the industrial chain being connected through.

7.4 Growth Analysis of Smart Learning Industry

There are three characteristics of the school learning system: government's overall consideration as the compass; market mechanism as the sally port, multi-participation as the overall context. Meanwhile, the industrialization of smart learning plays main force to accelerate the city development.

7.4.1 Industry Development Is Supported by Relevant Governmental Policies

With the in-depth advancement of the learning society, learning community, and learning city the government, become a powerful factor in promoting lifelong learning. Local governments and functional organization gradually increase their awareness, and establish specific measures and action plans promoting constant upgrade of the industrial layout of smart learning. At the national level, the policy of "Mass entrepreneurship and innovation" promotes the associated index such as knowledge, learning, innovation and happiness to be unified with living index; in the city level, various policies that are suitable for the young to start business, retain high-level intellectual talents and aim at "city and industry integration" are released in succession.

(1) **National policy: "Mass entrepreneurship and innovation" renovates and updates China's economy**. The "commercial registration system" reform carried out by the State Council, the series of policies for the reduction of reserve requirement ratio and interest oriented medium, small and micro-sized enterprises, and the measures that permit undergraduate students to be temporarily absent from school to start business, etc. to aim to lower the threshold of startup and upgrade entrepreneurial vitality. Teaching and learning industry, as the high-end smart industry, becomes one of the main directions for entrepreneurs. The implementation of "Mass entrepreneurship and innovation" results in two obvious changes: firstly, it promotes the renovation and update of economy; secondly, it promotes the expansion of learning and innovation market.

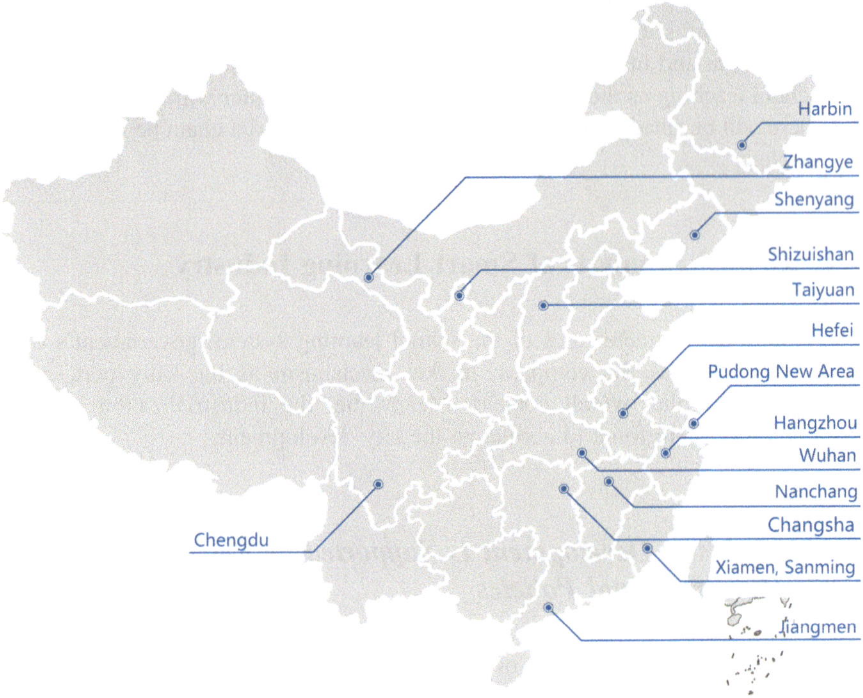

Fig. 7.22 Top 15 cities among the first group of "Mass entrepreneurship and innovation" demonstration

The Ministry of Finance, Ministry of Industry and Information Technology, Ministry of Science and Technology, Ministry of Commerce and State Administration for Industry and Commerce listed the top 15 cities as the first group of "Mass entrepreneurship and innovation demonstration" in June, 2015 (see Fig. 7.22). The "Mass entrepreneurship and innovation" demonstration cities put efforts to solve the outstanding problems of "the last mile". Focusing on the development of small and micro-sized enterprises, it explores how to establish the new mechanism by which the governments could support development of small and micro-sized enterprises.

(2) **Urban policy: the youth is the vital force for urban development**. The cities that are attractive for the youth possess two features. On the one hand, there is a good employment and entrepreneurial environment, which provides working and entrepreneurial opportunity. On the other hand, there is a relevant infrastructure and good social environment that meet the demands of the young for leisure, knowledge and recreation. The vitality of the city and the quantities of the young people living there promote mutually and interact with each other. Vibrant cities can retain the youth, and the young create value for the development of such cities and promote their upgrade.

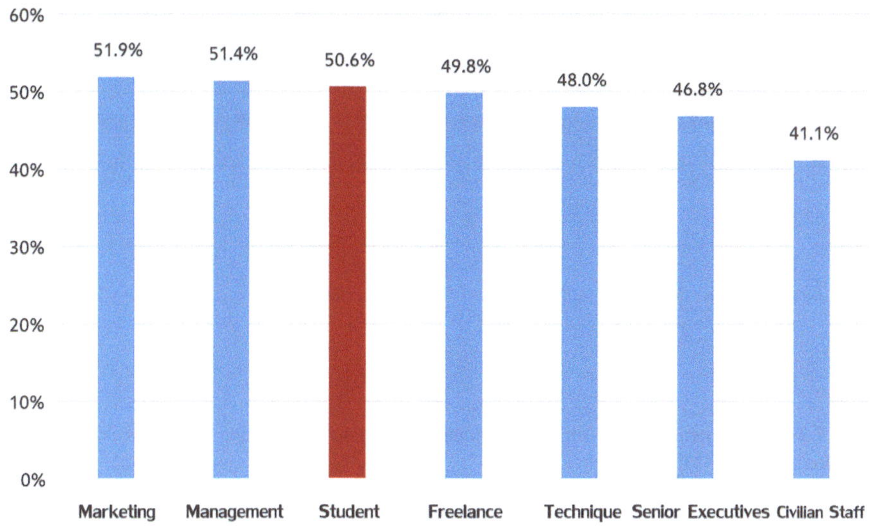

Fig. 7.23 Possibility of entrepreneurial potential of different population in China

Students are becoming the new main force for "dual-startup" of cities. According to *Entrepreneurial Potential Investigation of Population Segment in China*,[5] people aged 21–30 present larger entrepreneurial potential. Students rank the third in terms of entrepreneurial potential. These Internet natives will play a dominant role in forming new living and learning way in cities[6] (see Fig. 7.23).

(3) **Industrial policy: a large quantity of newly-established enterprises emerges**. Among the over 200 best sellers of educational application software in China, the sales of application software for preschool education takes up approximately 60%.

The newly-established enterprises of online education can be divided into: tool platform, flow platform, question bank, online preschool education, online assessment, curriculum schedule and terminal tool. In terms of the main audiences, the white collar employees with strong demands for improving vocational ability, and the students of K12 education, are still the mainstream groups (Sohu Education 2014) (see Fig. 7.24).

[5]Data source: Penguin Intelligence, *Entrepreneurial Potential Investigation of Population Segment in China*, Samples: 57375 pieces, January 2015.

[6]Data source: N = 43604. The ratio calculation: population with more entrepreneurial ideas in this profession/the total investigated number of people in this profession. Other options are excluded.

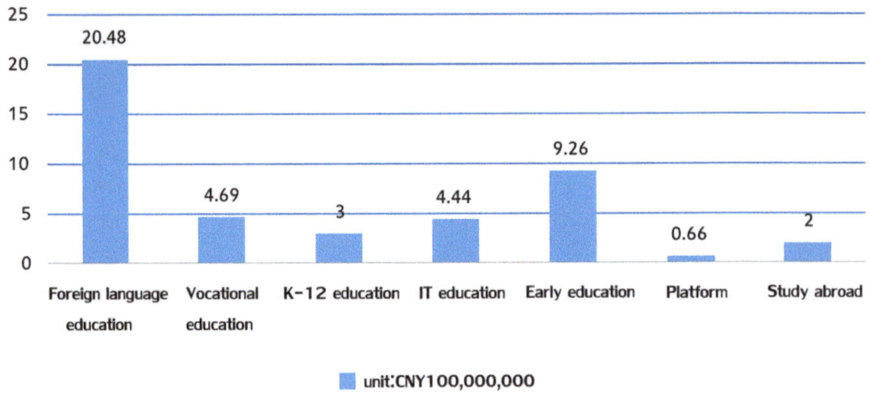

Fig. 7.24 Investment amount for the financing of each field of online education in 2014 (*unit* hundred million RMB)

7.4.2 *Smart Learning + Industry as a Key Indicator of a City's Soft Strength*

As the tool connecting members of information society, the Internet can effectively link teachers, students, parents and administrators. At the intelligent terminal level, such subjects can enjoy the learning environment at any time, place and in any pace. The structure principle of smart learning in a city includes 4 factors: intensive degree of cities' infrastructure, the asset-light coverage, the modern vocational educational system, and in-depth transformation of smart education and learning environment.

Principle I: Development of urban information infrastructure impacts the merits and demerits of smart learning environment. The input and layout of smart city in public educational facilities manifests the overall consideration of municipal administrators for digital soft power and city comprehensive competitiveness. The convenience, intelligence and efficiency of learning environment can be fully manifested in cities with perfect city infrastructure, including road traffic, residential buildings, environmental landscaping and educational facilities, etc.

Principle II: Light asset industry is a component of city industrial type. The epitome of city industrial value is the irradiation effect of high growth and high value-added industry. The teaching and learning industry belongs to light asset industry, fitting the innovation demand of the Internet. The future integration of urbanization, industrialization and Industry 4.0 will be manifested not only in economy and social security, but also in education and learning.

Smart city industry has reached each link of the industrial chain. Some actions of the state such as the Internet, Internet of Things (IoT), smart city construction, Internet + action and made in China 2025, etc. have been constantly creating new

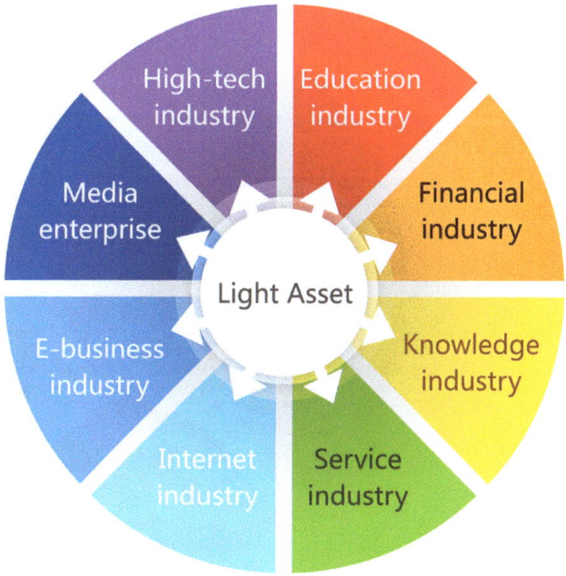

Fig. 7.25 Asset light industry types of city

types of business. Such new types of business will be overturned, divided and recombined into emerging ones in the general trend of the Internet thinking and cross-border integration. The emerging types of business dominated by asset light industry (National Vocational Classification Collection of the People's Republic of China 2015) are the beneficial supplement of manufacturing industry (see Fig. 7.25).

Principle III: Modern vocational education is the fresh force of city educational system. The rise of vocational education has become the essential trend. Firstly, labor force in city is structurally unbalanced. The ability to fit the professional development shall be enhanced through constant learning. Secondly, there are favorable policies as several draft amendments about educational laws were discussed and passed at the executive meeting of the State Council in January, 2015. The establishment at the level of laws for the commercial private education institutions will further promote the industrialization of education to become the normal status. In terms of the development of private education, various cities have issued encouraging policies in succession. Shenzhen has incorporated the development of private education into the overall planning of urban development (Shenzhen Special Zone Daily 2014).

Vocational education takes up 29.8% market shares in the overall Chinese Internet educational market (Analysys International 2015), what ranks it as a second after higher education. In this market, users have higher learning motivation, and stronger willingness to pay. Specifically, the enterprises of Internet vocational educational market are mainly divided into 6 types (see Fig. 7.26).

Fig. 7.26 Main types of internet vocational educational market

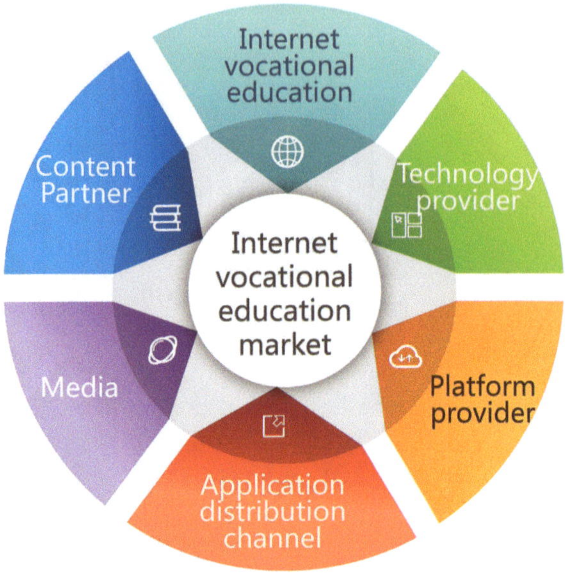

Principle IV: The learning environment of smart city needs in-depth transformation. The design of smart learning environment shall fully support the school education, vocational education, skill training and social cultural education. In the process of the rapid development of smart city, urban development will be impacted and limited to different extent by the lack of learning environment design, the improvement of citizens' moral quality and the overall image of urban brand. In the process of urban integration, the differences in learning environment among cities will in the future widen the competitiveness gap.

[Case] Shanghai: the forerunner of "MOOC" in China

Pudong New District of Shanghai is one of first experimental pilots for smart cities and is also the forerunner of "MOOC" in China. Shanghai deeply transforms the learning environment of urban residents mainly through 3 approaches. The first approach is Shanghai Micro-school, a large-scale smart learning platform for all learners. Each learner possesses a lifelong learning account. Various learning achievements of residents can be accumulated and converted; the second approach is course resource sharing platform of Shanghai colleges and universities, which gathers the power of the government, schools and enterprises. It is also faced with the bottleneck for interaction-interaction solutions; the third approach is educational data center, which offers help for educational scientific decision through the analysis on the learning demands, habit and effect of learners.

7.4.3 Industrialization of Production of Learning Content

(1) **Content is the key point to convert the traditional learning into smart learning**. According to *Online Education prospect and Hot Spot Analysis Report* (2013), the first factor decreasing the usage of online educational platform is "boring course content". To be welcomed by more learners, online education must improve the level of interaction between the user and the platform, what may increase enjoyment of course content and enhance the user's experience (see Fig. 7.27).
(2) **Different groups of people have different demands for the learning contents**. The content is the key to attract learners. Learners of different age have different demands regarding the content. What may appeal to students utilizing smart learning may include: instruction after class, interest and specialty learning, foreign language skill and computer skills. What may appeal to employees are: vocational skill, professional knowledge, various areas of interest and hobby, practical living skill, etc. Learners have imperious demands on superior contents[7] (see Fig. 7.28).
(3) **The development of learning content should consider the cross-platform operation**. It is in the transition period from E-Learning to M-Learning. PC and mobile equipment, as the output terminal of learning, will remain still in parallel state for the next few years. Meanwhile, there are will be new operating systems developed for the mobile terminals. However, when each operating system is developed for once respectively, the development cost will inevitably increase. The additional cost will be finally paid by the learners what will affect their selection preferences Therefore, industrial subjects need to think the "developing for once, using in multi-environment" for course contents.
(4) **The innovative teaching models are important for the development of contents**. Innovative teaching methods can break through the limitations of time and space, promote transformation of teaching models from the isolated ones to an open ones, and promote transformation of learning behavior from the inattentive one to a self-conscious one. There is also a need of transformation from the well-structured classroom teaching, to semi-open blending teaching, and to the fully open socialized teaching, that will change the structure of the relationship between teachers and students. For example, the flipped class constitutes a semi-open teaching and learning system; MOOC's character of networked and socialization manifests in-depth collaboration and open teaching.

[7]Data source: Netease Education and Youdao Dict (2013): Report on Online Education Trend in China in 2013–2014.

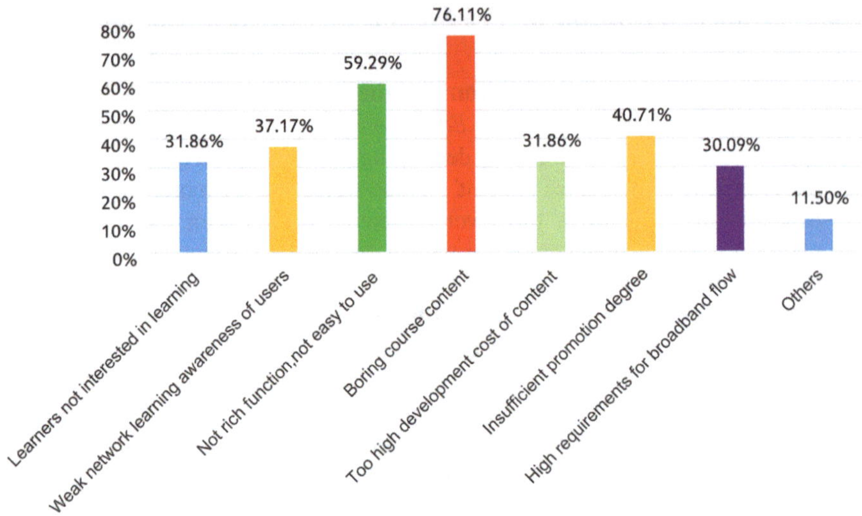

Fig. 7.27 Influence factor on the usage of online educational platform

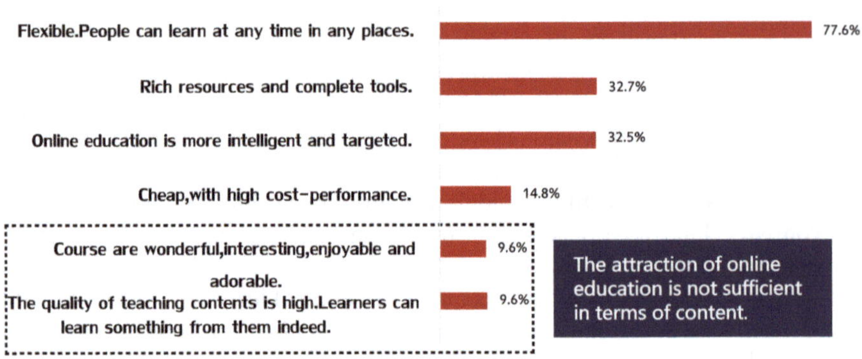

Fig. 7.28 Content insufficiency of online education

(5) **The innovative model of smart learning utilizes superior learning resources**. It equips learners with dynamic self-development ability through use of advanced science and technology. For example, the "smart information factory mode" of Google search engine can solve the problems that cannot be settled by individuals. This allows learning contents be spread more broadly.

7.4.4 Evolution and Reconstruction of the Industry Chain

The layout of the overall smart learning industrial has been already started. It is not only the key industry that the Internet seeks to overturn, but also the mainstream of the new economic entity in the future. **Smart learning industry is involved in the two aspects of traditional education: industrial chain and socialization learning industrial chain**. Their rise will not only impact the traditional education industry, but also result in new industrial models. The current integration period is the stage that various parties scramble for market positions.

(1) **Traditional education industrial chain enriches the connotation of city industry**. Taking "government-school-family" as the subjects, the government-dominated traditional education industrial chain is in an urgent demand for new elements, new concepts and new markets. The B2B2C closed-loop ecology of education industrial chain is formed rapidly, which is manifested in two aspects: firstly, due to the openness of industrial field and industrialization trend, increased number of participants, and market segmentation are obvious. Original resources are fully vitalized. The industrial pattern is formed, taking the industrial capital-upstream-midstream-downstream as the bond; secondly, taking students as the center, the industrial chain expands laterally, with innovative capital, integrated resources, extended channels and rich contents. It covers each level of learning subjects, meets the lifelong learning demands and riches the connotation of city industry (see Table 7.3 Education Industrial Chain).
(2) **Education industrial chain is gradually integrated with the socialized industrial chain**. The city-class education industrial chain is at the key period of exploration, integration, innovation and collaboration. At the capital level, it is mainly manifested by two aspects: capital management and capital cooperation. At the upstream of industry, it promotes the conversion rate of education resource through the integration and creation of resources. It also results in innovative supporting services. At the midstream of industry, it is pivotal to connect the provider and demander of education; At the downstream of industry, the number of demanding learners is constantly increasing, what in the future will promote the upgrade and transformation of the industry.
(3) **The family, society and government jointly upgrade the industrial environment of smart learning**. Industrial environment and industrial pattern promote mutually. Firstly, the government, school and family jointly create an open and comprehensive industrial environment. Secondly, technology, market and enterprise form the joint force to promote the development of industry Thirdly, the self-innovation of industrial environment and the active adaption of industrial pattern promote each other, and jointly purify the smart learning environment (Fig. 7.29).
(4) **Three models reconstitute the industrial pattern of education**. Before the rise of smart education, the attraction and profitability of Chinese education industry decreased as a result of wrong industrial policy and deficiency of

Table 7.3 Education industrial chain

Industrial capital		Industrial upstream		Industrial midstream		Industrial downstream	
Capital governance	Stock, financing, investment, management and withdrawal …	Resource integration and creation	Site equipment, Material items, participants, resources in system, products, clients …	Resource aggregation and utilization	School center, tutorial institution, online platform, Expo procurement …	Institution	Government institution, traditional education, Internet enterprise, vertical platform, research association …
Capital cooperation	PPP Government + school, School + society, Society + society …	Supporting service innovation	Technical platform, content channel, information consultation, added value …	Alignment of upstream and downstream	Investigation planning, consulting evaluation, e-business enterprise, added value …	Individual	Students' parents, workers, specialists and scholars, the senior and children …

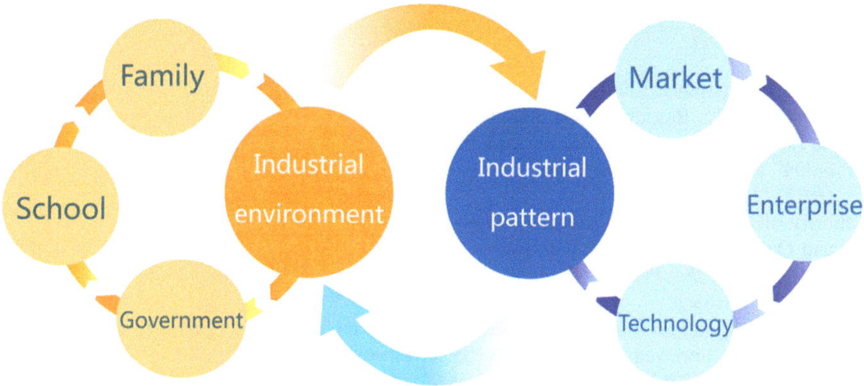

Fig. 7.29 The role of family, society and government in facilitation of smart learning

business model. Since the rise of smart education, the education market has been gradually open. Accordingly, three learner-centered mainstream industrial models emerged: traditional model dominated by ICT in education, Internet model dominated by Internet enterprise cross-border, vertical model dominated by professional cyber-schools. They lay the foundation of market-oriented industrial development pattern (Fig. 7.30).

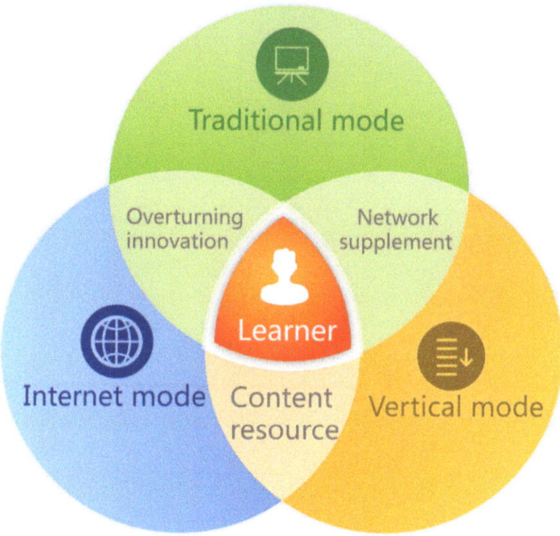

Fig. 7.30 Relationship of the three models

References

Analysys International. (2015). *Mono-graphic study on internet vocational education market in China in 2015*. Retrieved from http://www.wtoutiao.com/p/gccInD.html

China Economy Network., & Mobile Learning Information Network. (2013). *Analysis report on online education prospect and hot spot*. Retrieved from http://edu.ce.cn/zg/201311/18/t20131118_1212477.shtml

Ministry of Labor and Social Security, State Administration for Quality Supervision, Inspection and Quarantine., & National Bureau of Statistics. (2015). *National vocational classification collection of the People's Republic of China*. Beijing: China Labor & Social Security Publishing House.

Netease Education., & Youdao Dict. (2013). *Report on online education trend in China in 2013–2014*. Retrieved from http://edu.163.com/special/onlineedu/#p=9G7TEOLS3Q0N0029

Power Education. (2014). *Prospect on profit mode of K12 online education*. Retrieved from http://edu.qq.com/a/20140702/022128.htm

Sohu Education. (2014). *White paper of education industry in China in 2014*. Retrieved from http://learning.sohu.com/20141202/n406594224.shtml

Wang, Q. (2014). *Understanding product form of domestic online e-ducation via one article*. Retrieved from http://edu.qq.com/a/20140421/012950.htm

ZHIYAN consulting. (2013). *Investigation and investment prospect report on preschool education development state in China in 2013–2018*. Retrieved from http://www.chyxx.com/research/201309/219302.html

Zhou, Y. C. (2014, January 2). Promoting the development of private education in Shenzhen through multiple measures. *Shenzhen Special Zone Daily*, P.A03.

Part III
Future Trends: Chinese and Multicultural Perspectives

Chapter 8
Future Trends of Smart Learning: Chinese Perspective

Abstract This chapter outlines the developmental trends of smart learning in China, and offers suggestions for the government how to promote smart learning industry in the long term. Firstly, the author discusses smart learning environment in China from the perspective of research and practice, as well as current changes in educational system. The impact of information revolution on smart learning is analyzed from the perspective of learning society, learning behavior, mobile learning, smart city and learning innovation. Finally, the future trends of smart learning are discussed in the context of integration with "Internet+", reconstructing the structural of learning, reconstructing learning environments, reforming online education, and upgrading smart learning. The chapter closes with suggestions how to further promote the development of smart learning.

Keywords Future trends · Policy support · Financial input · Technology innovation · Development suggestions

8.1 Introduction

In China, the improvements of smart learning environments could be attributed to the following factors: the rapid pace of innovation in information technologies, the growth of academic attentions, the increasing financial input, and the rising importance of smart learning has for the local governments constructing smart cities. The broadly accepted by governments concepts "Knowledge Economy" and "Internet+" have redefined the word "learning" and its connotation.

By applying information technologies such as Cloud Computing, Big Data, Mobile Web, Internet of Things, etc., smart learning brings several significant changes for the society. It eliminates boundaries between formal education, family education and social education, and transcends the limits of time and space. Smart learning platforms allow learning take place not only on the campus, but in the broader social context, and provide chances for everyone to access high quality education resources. Smart learning tools, mainly mobile applications, create

D. Liu et al., *Smart Learning in Smart Cities*, Lecture Notes in Educational Technology, DOI 10.1007/978-981-10-4343-7_8

interactive learning scenarios that mobilize improve people to learning, and encourage them to make use of every small pieces of time to develop their knowledge and skills.

Meanwhile, people need to develop the capability to screen true knowledge from the huge, decentralized web of information (Simons 2009). In the near future, with the development of technologies such as artificial intelligence, machine learning, virtual reality, augmented reality, etc., customized smart learning tools might be developed, and the learning scenario will be more real and vivid. The smart learning helps to bridge the gap of different regions and cities. It is also benefits to the disabled group (Ministry of Education 2015).

The Chinese government advocates to build a learning society and smart learning is an important part of it. The government has introduced policies and increased financial support to development of smart learning industry. This chapter provides some suggestions for the government to promote the smart learning industry and environment.

8.2 Environments for Smart Learning

The evolution of digital learning environment towards smart learning environment is based on the integral compatibility of smart city development, technical progress and universal benefits of information technology. Smart learning is not an isolated system. It includes not only the digitization of formal education systems in structured learning environments, but also the open learning systems faced to families and the society (Huang 2014). The prospect of smart learning intensively embodies the purpose of smart city and smart education. Knowing the past, present and future of smart learning is the precondition for understanding the practices of smart learning nowadays.

8.2.1 Research on Smart Learning Grows Fast

There are two different paths about the research on smart learning in Chinese academic circles. One is based on the digitization of education, with particular emphasis on the innovation of formal and structured education; the other is based on smart city, with particular emphasis on the openness of education and inclusiveness of education resources.

Searching in the database of China National Knowledge Infrastructure (CNKI), the academic trend diagram of "digital learning, online learning, smart education and smart learning" was shown in Fig. 8.1. It can be seen that, digital learning, smart education and smart learning are attracting more attention from scholars in recent years.

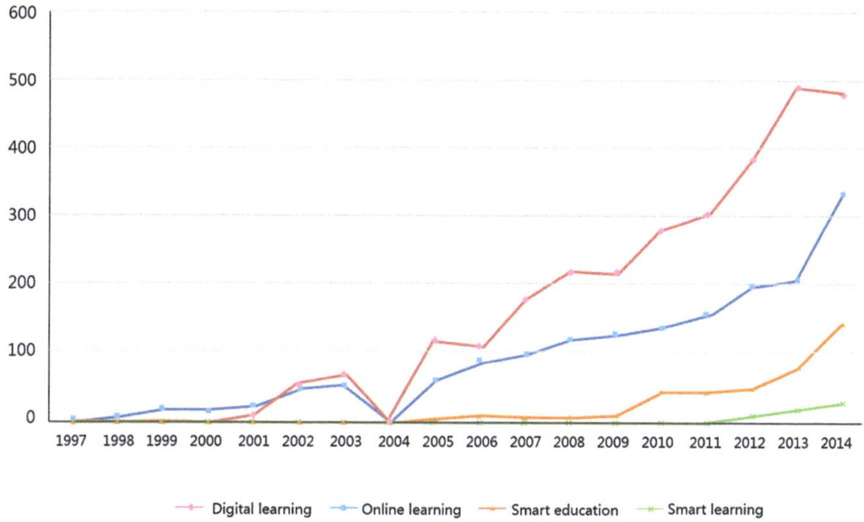

Fig. 8.1 Academic trend of digital learning, online learning, smart education and smart learning

The difference between smart learning and traditional learning is that smart learning is a brand new type of learning based on digitization, collaborative innovation and knowledge fusion, while traditional learning is the learning at classroom with books and blackboard. With the support of "information factory + Internet + big data", smart learning breaks through the difficulties of the current education, through the key point of transforming teacher-centered education to learner-centered education.

The number of research institutions of smart learning is constantly increasing. The institutions of higher education take the dominant position for research on smart learning. These are the three main growing trends of the development of smart learning research organizations in China. Firstly, taking China Wisdom Engineering Association (founded in 2005) as the criterion, research institutions of comparable size or level emerge rapidly. Secondly, the developing areas have the same amount of innovation initiatives as the developed areas in carrying out research on smart learning, with consideration of practical local conditions. Thirdly, the blended model of association between schools and society as the research entity is adopted by most of the smart learning research institutions (see Table 8.1).

It is believed that the rapid emergence of such research organizations is mainly attributed to three facts. Firstly, the standpoint that smart learning will be the main type of learning in the future has become the social consensus. Secondly, compared with traditional learning, smart learning helps to reinforce human resources which will support regional economic growth in the long run, narrowing gaps between developed and less developed regions. Thirdly, smart learning, which carry out the basic research of smart city, smart industry, smart science and other smart industrial chains, could promote the combination of production, teaching, scientific research and application.

Table 8.1 Some institutions carrying out the research on smart learning

Name of institution	Founded time	Initiating unit	Mission
China Wisdom Engineering Association	In 2015	Initiated jointly by national key scientific research universities such as Beijing University and Tsinghua University, etc., scientific research units and celebrities in all sectors of the society	To research smart science, cultivate smart talents, and improve the smart level and innovation ability of Chinese nation
Beijing Smart Siyuan Institute	In 2009	Tsinghua University, Beijing University, Beijing Normal University, etc.	To constitute the multi-education system of school education, family education and social education
China Mobile Learning Alliance	In 2014	25 initiating units such as Open University of China, *Distance Education in China*	To serve the development of mobile learning
Smart Learning Institute of Beijing Normal University	In 2005	Beijing Normal University, NetDragon Websoft	To promote the two-way fusion of information technology and education, and support digitization of education in China serve the education digitization construction of China

[Case] West Lake Reading Festival

Since 2007, 8 West Lake Reading Festivals are held in Hangzhou in total. Each reading festival lasts for several months, with hundreds of activities and millions of participants. In addition, according to the data of China's mobile phone reading base, the users reading through China's mobile phone are up to 3.06 million in Hangzhou, ranking the second among all the cities in China.

8.2.2 Practices of Smart Learning Grow with the Increasing Educational Financial Input

The input intensity in smart learning is an important indicator of the sustainable development of smart learning. Contemporarily cities in China are investing largely in improving both hard and soft environment of human settlement. With the systematic implementation of urban integration, smart learning is no longer limited in

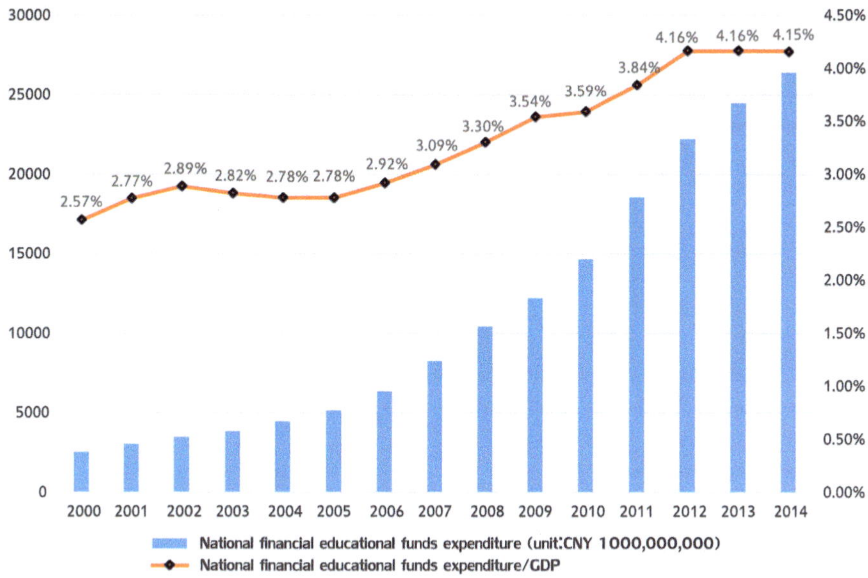

Fig. 8.2 Growth of financial educational expenditure in China in 2000–2014

the single educational filed. Learning itself becomes the important symbol for person to merge into urban development, adapt to urban change and improve grade.

(1) **The proportion of financial investment in GDP is up to 4%, reaching the world level baseline**. The *Outline for Reform and Development of Education in China* printed by the Central Committee of the Communist Party of China and the State Council in (1996) put forward that the proportion of national financial education expenditure in GNP should reach 4% in the end of the 20th century. However, public notice on the educational expenditure statistics in China in 2012 indicated that the proportion of educational input in 2012 was 4.28%, which just met the standard. This change can be seen from the input over the years[1] (see Fig. 8.2).

The realization about the 4% target results from the three supports of policy, finance and local region, which is obtained through multiple years. It is not the terminal, but a starting point. The future knowledge structure and information society are full of the competition of talents. There are still many weak links in education among cities. The proportion of educational input remains to be further enlarged.

(2) **K-12 education and online education received more fund from government and industry**. The industrial input strategies in China lay emphasis on the extension to the upstream (Capital Laboratory 2015) (see Fig. 8.3). The highest

[1]Data resource: *Statistics Bulletin on Education Expenditure Execution Conditions* of Ministry of Education over the years, statistical yearbook of State Statistics Bureau, 2000–2014.

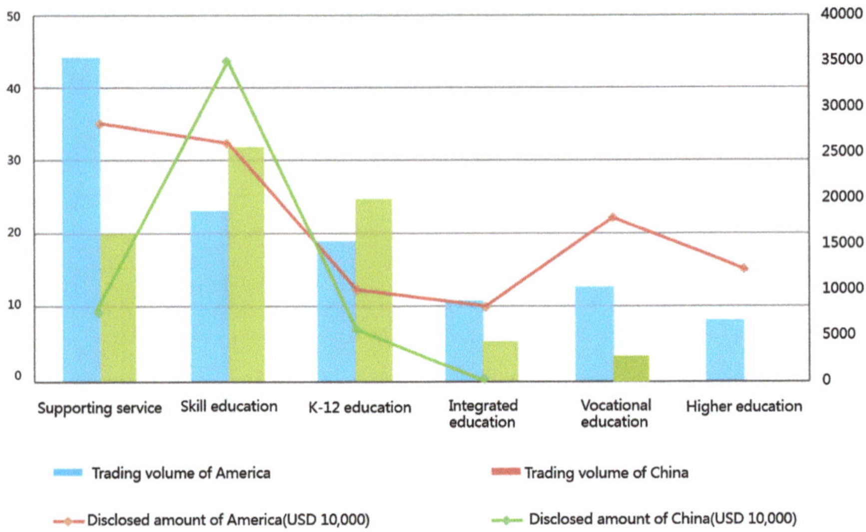

Fig. 8.3 Venture investment distribution for online education in China and America in 2014

amount of investment in online education in China is centralized on skill education, much higher than that of the supporting services. It indicates that online learners want to learn some practical skills.

Online education is conductive to making up the gap of educational resources among cities. The first-tier cities gather high-quality educational resources, superior teaching staff and better learning environments. In addition, it becomes the most preferred resource place for online education learners. The *Internet Learning User Behavior Report* issued by HuJiang Online Education (2005) indicated that the growth rate of students in Beijing, Shanghai and Guangzhou is lower than the conventional tier-2–tier-4 cities.[2] In terms of the daily learning frequency, the annual growth rate of the above-mentioned areas is also much higher than that of Beijing, Shanghai and Guangzhou.[3] The main reasons lie in the lack of traditional educational resources in tier-2–tier-4 cities. To obtain high-quality learning resources, learners turn to seek for online education. The high-quality educational resources of the tire-1 and tire-2 cities flow to the training classes of the tire-3 and tire-4 cities, which realize the spread and utilization with lower cost and larger scale.

[2]The city tiers in China mainly depends on the population, income, and competitiveness. The first tier cities are the four biggest cities in China: Beijing, Shanghai, Guangzhou, Shenzhen; while the second tier cities are mainly the capitals of each province.

[3]Report on Internet Learning Users' Behavior issued by hjclass.com. [EB/OL]. http://mt.sohu.com/20150528/n414011755.shtml.2015-5-28.

8.2.3 The Scope of "Learning" Changes in Knowledge Society

Knowledge Economy breaks through the limitation of campus learning, creating a school without any bounding walls. The rise of various informal forms of learning enlarges the selection scope of learners, enlarges the extension and enriches the intension of learning. The society is entering an era of open, socialized and ubiquitous learning.

(1) **Learners can be found not only on campuses, but in society at large**. Compared with full-time students, regular citizens have less autonomy in both learning and acquiring learning resources, and it is not easy for them to get access to high quality educational resources. In the congratulatory letter of international conference on ICT in education in 2015, President Xi Jinping urged to establish a "anyone, anywhere, any time to learn" learning society. More citizens and people in remote areas get the opportunities to learn what the they want through quality online learning resources. The learners are in the network-like distribution: a large population, wide distribution, complex structure, uneven levels.

Two different statistics from the same period, indicate that the growth rate of the Internet learners is continuously accelerating. The research report jointly issued by Baidu Library and Ministry of Education (2014) shows that there were 340 million Internet learners up to October, 2014, with the month growth rate 7%. The volume was approximately 390 million up to December, while the 35th report of CNNIC in the same period (2015) indicated that the quantity of the Internet users was 649 million. The Internet learners took up more than half of the Internet users, as shown in Fig. 8.4.

Learning is not only the transfer of knowledge, but also the upgrading of skills. The online learning helps learners transform fragmentary knowledge into skills. There are mainly two groups of learners that should be focused in online learning.

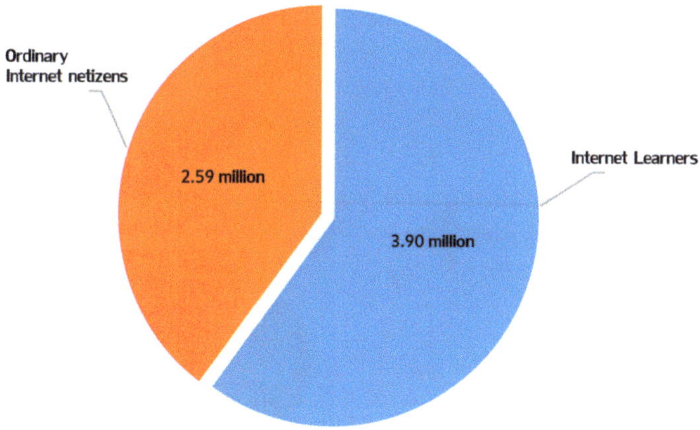

Fig. 8.4 Quantity contrast between common internet netizens the internet learners up to December, 2014

The first group is those who live in remote areas with scanty educational resources (e.g., countryside, the west areas, etc.). The second group is those of migrant workers in the process of China's industrialization and urbanization, who merely benefit from structured education and have little vocational skills, and need adapt to the urban life and the pace of work through continuing education.

(2) **The subjects of evaluation are diversified**. The single evaluation model is a structural tower layer model. Teachers and educational institutions are the subjects of evaluation. The multi-evaluation model is a de-centered net structure, and students are the core of the evaluation system. Teachers, educational institutions, parents and the public take part in the evaluation for student's learning (see Fig. 8.5). There are three reasons for the diversity of evaluation. Firstly, diversified evaluation subjects can enrich the evaluation system; secondly, it promotes the "teaching and learning" benign interaction; thirdly, it promotes economic sharing through learning sharing. Sharing economy is the new approach to stimulate economic growth.

Generally, the normal conditions of learning include three steps. The first step is evaluation from diversified subjects, which evaluates the level and learning status of learners and inform the results to teachers or students; the second step is the learning process, in which learners interact with peers, teachers and contents; the third step is re-evaluation, which evaluate the learning effect. In these process, learners can obtain the evaluation and feedback about learning in time, and adjust learning direction and make up the deficiency of personal knowledge system.

(3) **Community education, family education, and special education are becoming more important**. The purpose for the development of national education is to establish a learning society for civil-learning, lifelong-learning, and any-time/pace-to-learn. In the network age, the learning society is gradually formed with the acceleration of the Internet. Community education, family education and special education become more and more important (see Fig. 8.6). Family education is the bond between school learning system and city learning system. Without a

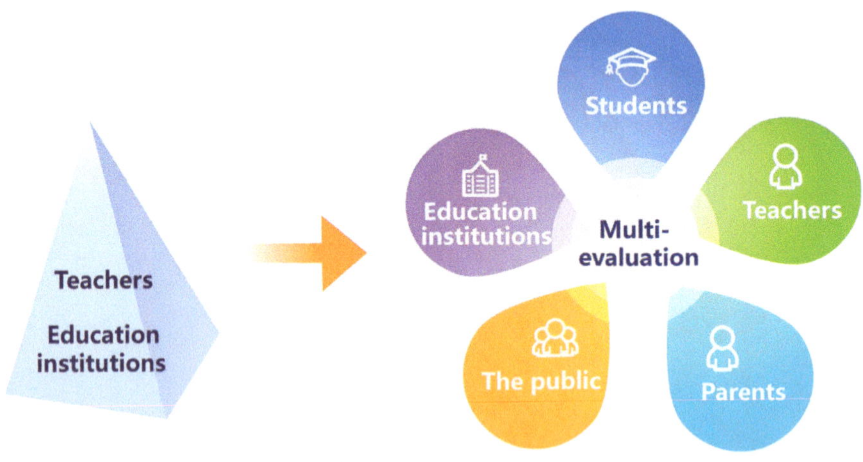

Fig. 8.5 Evaluation subjects developing from singleness to diversity

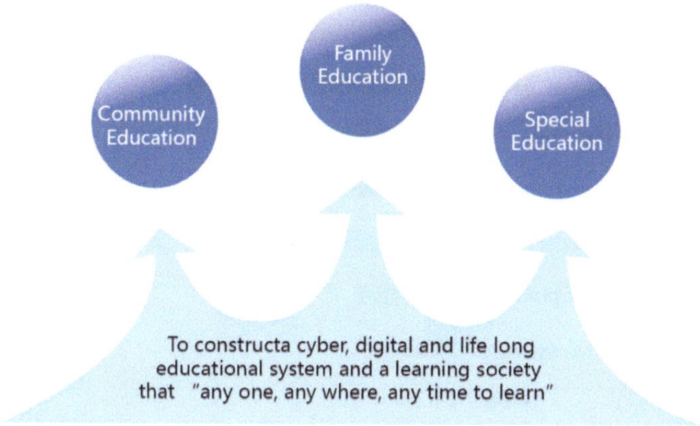

Fig. 8.6 Subjects of three types of implicit education becoming explicit Gradually

good family education, faults will appear between the school learning system and city learning system.

The rise of social media forms community organizations and community economy. Community education is an important component of social organization, the cell of city education, also an effective channel to form good lifelong learning atmosphere; family education is the bridge to connect school education and social learning. Special education is an important part in education system. There are over 2000 special education schools in China at present, with 400,000 undergraduates in schools (Ministry of Education 2015). In terms of courses, textbooks, teaching methods and teaching modes, etc., the "on-demand education" that meets special demands combines the special education, community education and family education, and protrudes the broad education trend. Taking family education as an example, according to *Chinese Family Educational Consumer Map in 2015* issued by the largest domestic parents community "jzb.com", the families with higher income have a larger proportion of education expenditure, and prefer network media in terms of channel selection[4] (see Fig. 8.7).

Internet is conductive to improving citizens' scientific literacy. Internet can promote the spread of scientific knowledge. The results of the 9th Chinese Citizens' Scientific Literacy Investigation in 2015 indicated that the proportion of the citizens possessing scientific literacy was 3.27% in 2010, and reached 6.20% in 2015; 91.2% of them obtained scientific and technical information through Internet and mobile Internet.[5] Internet is the first channel for the citizens with scientific literacy to obtain scientific and technical information.

[4]Source of data and figure: jzb.com and Future Education Institute; January 5, 2015–January 12, 2015, Online Investigation Data, N = 1741.

[5]The 9th scientific quality investigation results of Chinese citizens issued by China Association for Science and Technology [EB/OL].

http://education.news.cn/2015-09/19/c_128247007.htm.2015-9-19.

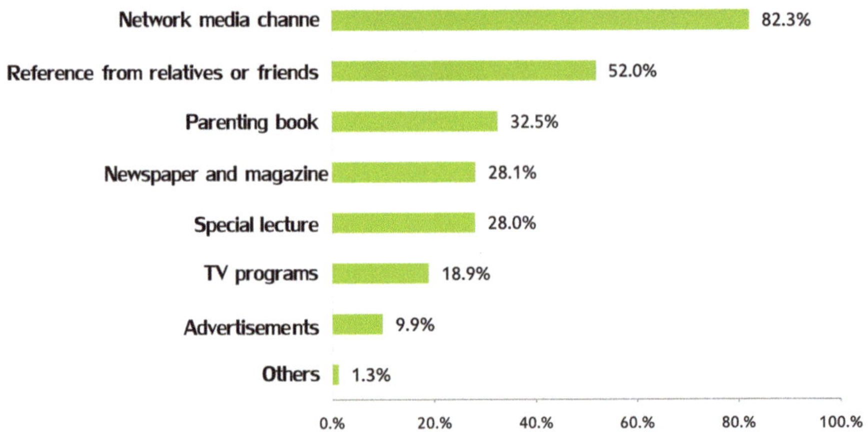

Fig. 8.7 Channels of Chinese parent (family) education to obtain education information in 2015

(4) **Reading is conductive to adapting to the changing times**. After entering the information society and knowledge economy era, reading becomes one of the most important learning abilities. State Administration of Press, Publication, Radio, Film, and Television of the People's Republic of China issued *Notice on Carrying out 2014 Citizen Reading Activity* in March, 2014, which encouraged each region to positively promote the nationwide reading legislation to direct nationwide reading into the orbit of legislation. Serious reading is the way that must be relied on for the inheritance of human knowledge. With the progress of technology, the range of reading carriers is extending gradually. With the increasing popularity of electronic reading, fragmentation reading and paper reading will run parallel in a long term in the future. The results of 12th nationwide studies on citizen reading indicated that the contact rate of digital reading mode of Chinese citizens shows an upward tendency[6] (see Fig. 8.8).

The reading volume of juveniles is increasing in general and is higher than that of adult citizens. The group aged at 14–17 possesses the largest reading volume. The reading volume of the children aged at 0–8 is stable—approximately 5 books (see Fig. 8.9); among the families of the children aged <8 with reading behavior, the families that have the habit to reading together with children at ordinary times take up 86.5%. Such families take 23.87 min to accompany children on reading every day on the average, which promotes the improvement of Chinese citizens' reading level in the feature.[7]

[6]Data source: Focusing on Reading Investigation of National Citizens: Mono-graphic Investigation on the Reading of National Citizens [EB/OL]. http://www.chuban.cc/ztjj/yddc/.

[7]Data source: Focusing on Reading Investigation of National Citizens: Mono-graphic Investigation on the Reading of National Citizens [EB/OL]. http://www.chuban.cc/ztjj/yddc/.

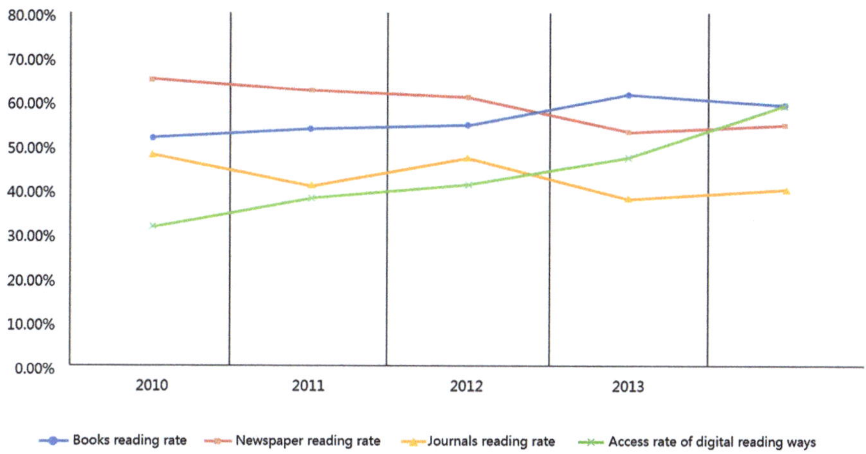

Fig. 8.8 Comprehensive reading rate of various media in 2010–2014

Fig. 8.9 Reading volume contrast between juveniles and adults in 2010–2014

8.2.4 Education Becomes the High-End Service in Cities

High-end services are the main force for urban development. During the period of "the 12th Five-Year Plan", the growth of services is faster than that of industry. It became the biggest industry in China in 2013. According to the exploration in Baidu Encyclopedia, the high-end services are intelligent, capitalized, professional and efficient services. It gradually influences the future of cities. The educational industry is an important category in urban high-end services, which has two values to city development.

Value I: Educational industry driving the transformation and upgrade of urban services. The educational industry is a high-end industry with high added value and driving force. In the network age, it is conductive to forming the soft power of cities and building a new open pattern of cities. For example, Beijing lists educational industry as one of the six major fields.

[Case] Beijing giving full play to comparative advantages

Beijing lists cultural educational services as a major field, at an equal pace with scientific and technical service, Internet and information service, financial service, business and tourism service, and health and medical service, for collaborative development.

In terms of content: clear to enlarge educational open cooperation, encourage foreign investment to establish the schools for the children of foreign personnel, support foreign enterprises to invest and establish educational training institutions and projects in the way of Sino-foreign cooperation in running-school, actively introduce world-famous schools for Sino-foreign cooperation in running-school, and to realize the benign interaction of educational resources.

In terms of auxiliary supporting system: define the talent aggregation mechanism, increase the overseas high-level talent introduction intensity of six major fields, and rapidly handle multi-entry visa or residence permit; open a green channel for the application for permanent residence of high-level talents; and encourage high-level talents to start business in Beijing, etc.

Value II: Satisfaction level of education is an important indicator to inspect the happiness level of urban citizens. In the framework of "livable and innovative" dual-core framework of smart city, the 21st century skills, inclusive education and ICT into education, etc. are conductive to urban purification through the lead of culture. The "CCTV (Central China Television) Economic Life Investigation" sponsored by CCTV China Business News indicates that: the groups with higher educational level possess stronger edit the sentence, not clear sense of happiness. Knowledge perfects personality, opens the door to wisdom, and is conductive to improving the meaning of life and creating value. Therefore, the first task to enhance the livable index and happiness level of cities is to intensify efforts to improve urban educational situations.

8.3 Influence of Information Revolution on Smart Learning

The information revolution establishes a ecological system that linked human, behavior, value and technology. In this system, the core is human who is served by technology. The value of Internet technology is to popularize education and learning to more people in a more convenient way. This path is from bottom to top, to help learners improve learning environment via technology, and grasp the learning and survival mode that adapt to the pace of times, so as to promote the corresponding changes of courses, teachers, schools and policies. The influence of information revolution on smart learning could be illustrated from the following aspects.

8.3.1 "Internet+" Contributes to the Learning Society in Cities

In the process to promote the transformation and upgrade of learning society, "Internet+" can optimize the learning environment and result in richer learning experience, more matched learning resources and more timely learning feedback. It propels the promotion of learning industry and the improvement of urban learning behavior.

(1) **"Internet+" action plan gradually covers education and learning society level**. Internet+, as a new economic form, expedites new patterns and new opportunities, and promotes the in-depth integration of innovation achievements into each economic and social field. After the "Internet+" action was put forward in the government work report at "NPC & CPPCC" in 2015, the "Internet + Action" plans have been issued intensively across the country. The regions in plan incorporate smart education into key special projects. Some regions have mentioned to promote mobile education. Up to now, approximately 31 provinces and cities have mentioned the concepts and measures associated with education[8] (see Fig. 8.10).

Education resources are socialized. Shanghai takes the establishment of city-class education resource center as the sally port, to promote high-quality education resources. It supports the education management decision, teaching research and public information service via big data. Foshan establishes the "digital learning harbor" through the innovation and development of Internet education service, and builds multi-level multi-channel citizen-independent learning network and learning platform. Hangzhou serves the construction of smart city and learning city through high-quality education resources, and improves education quality of the city.

Learning services are platform-based. Lifelong learning is incorporated into most "Internet + Action" plans. It makes learning services and citizens' development integrated with urban construction through the paths of service platform and cloud-based platform. Dalian focuses on digital libraries, digital museums, digital archives and E-reading rooms, to promote the construction cultural information resources sharing project; Qingdao is dedicated to exploring the educational and cultural smart service system oriented to requirements of the society. Fujian

[8]Data source: White Paper project team, statistical data of "Internet + Action" in various regions.

Fig. 8.10 "Internet + Action Plan" distributing smart learning

establishes the strategy that is "platform + application + terminal + content" and network education preferentially developed.

(2) **The establishment of learning city is the foundation for building a learning society**. Up to now, nearly a hundred prefecture-level cities in China has set the goal to build learning cities and has been actively implementing this goal of constituting learning city and have been practiced them actively; seven ministries and commissions such as Ministry of Education jointly issued *Comments on Promoting the Construction of Learning City* in 2014, which future promoted the fulfillment of the learning city. It has laid foundation of the learning society through the overall shaping for it. The construction of learning city has important practical significance as it promotes the lifelong learning and the all-round development of people.

[Case] Hangzhou as the model of learning city

Arne Carlsen, the Director of Lifelong Learning Institute of UNESCO evaluates that: Hangzhou is a vivid case for the construction of global learning city. Its experience is worthy of active promotion. The experiences of Hangzhou mainly include.

It has basically formed the public cultural service system covering the four levels of city, district or county (county-level city), street (town), community (village). Currently, there are 20 libraries, 84 museums, 90 theaters, 190 street (town) cultural stations, 42 activity centers and 40 cultural squares. The quantity of infrastructure comes out among the best of the same kind of cities in China.

Hangzhou listed the Hangzhou Regulations on Lifelong Learning Promotion listed in the legislative research project in 2015. Combining the construction of smart city, it promotes the online learning by taking the advantages that operators of mobile phone reading platform are settled in Beijing; it advocates and popularizes the learning models such as MOOC and micro class, etc., to create a learning society that takes lifelong education system and lifelong learning service system as two pillars and takes innovative citizen learning activities as the carrier.

(3) **The youth is the fresh force of Internet natives and learning society**. One of the challenges faced by the cities is aging of the population, the other inability to remain attractive for young people. Challenge faced by cities, aging of population and insufficient attraction to retain the young is a prominent problem. In the Internet world, the generations after 1990s and 2000s are called as Internet Natives.

One peculiarity of the Internet Natives is the close relation between living and learning behavior and the Internet. There is both consistency and significant difference between the learning of learning society and the traditional learning. It requires us to break through the limitations of isolated and infusing learning, and conduct organized and socialized interactive learning via modern information conditions. As such, the traditional learning at school only for teenagers turns to learn in the whole life span. This learning way upgrades the rational cognition and optimistic judgment to the future of China by Internet Natives. They obtain more comprehensive information through the Internet. It is conductive to analyzing the current status and future of China's economy. According to the *Network Society Mentality Report of China* (*2014*) issued by Fudan University (2014), the generation after 1990s is a relaxed and optimistic "social generation", a group with the most optimistic mind about the development of China.

The key to the innovative development of cities is to retain the young generation after 1990s. During the revolutionary transformation of all walks of life through the Internet, it is creating the new engine of economic growth, to retain "people" in cities and make them live and work in peace and contentment.

8.3.2 ICT Reconstructs Learning Behaviors

Different learners have different requirements for learning environment. Smart learning environment can be divided into four categories supporting: "self-study"; the "researching learning"; "learning at work"; and "learning by doing" (Huang et al. 2012). Such environments have a distinct feature: learner-centered. In addition, they reconstruct learning behavior from the four aspects of urban system, knowledge system, environmental system and cognitive system.

(1) **Urban system: intellectual laborers are the cells of smart learning**. In the context of dual-startup, intellectual, skilled and innovative laborer force[9] can collaboratively complete the complex work on various intelligent platforms. Meanwhile, they establish horizontal, learning-oriented and fast-response smart organizations: smart enterprises, smart governments, government communities and government's educational institutions, etc.

According to international experiences, social consumption structure undergoes great changes when per capital GDP exceeds USD 1000. People have greater aspiration to pursue rich life contents, improve life quality and improve life environment. The per capital GDP in China has already exceeded this phase. The initiation of smart cities realized object-object, object-man, man-man interconnection and interworking and environmental awareness, which is the foundation for information consumption. The tremendous learning and innovative potential of people are activated in multiple ways. To accomplish this smart cities, we need to establish and improve the problem-based and learner-centered systematic learning management mechanism.

(2) **Knowledge system: the self-constructed knowledge system is superior to the system constructed externally**. The learners in network age need to acquire the ability to search, find and analyze knowledge in the ocean of information. The learning in the network era is de-centered. Knowledge is no longer provided by authoritative scholars, instead, a huge quantity of information and knowledge fragments is provided by lots of netizens (Simons 2009). Therefore, knowledge formed in the network era have much differences with knowledge formed the industrial era (see Table 8.2).

(3) **Environmental system: smart learning environment is formed via the connection of Online to Offline (O2O)**. O2O is a business strategy that draws potential customers from online channels to physical stores. It has evolved as one manifestation of the Internet era, with its influence in the education and learning level. Online learning platform provides basic information and some training for learners, and offline service platform provides physical service to learners (see Fig. 8.11). O2O will reshape the learning concept and learning model.

Online learning helps learners develop lifelong learning ability. Learner-centered learning emphasizes learning motivation and assists learners in make planning of their lives. Due to the pressure from the company and society most people loose motivation to acquire new knowledge when their career begins. The fragmentation of learning can be implemented by making the most of fragmentary time; educational resources are always online, so learners can cultivate the lifelong learning abilities.

[9]Source: Xi Jinping's address at the "May 1" International Labor Day celebration and the national model workers and advanced workers commendation conference (2015).

Table 8.2 Knowledge system in different ages

Times	Structure of knowledge	Construction mode	Learning behavior	Learning system
Industrial age	Systematic	Teachers	Guided by teachers	Traditional learners, tower-layer knowledge system
Information age	Intact and systematic	Others or institutions	Reassembled in the network	Online learners, re-centered knowledge
Network age	Scattered and fragmentary	Self-constructed	Reconstructing knowledge system by themselves	Self-dominated, spider-web knowledge structure

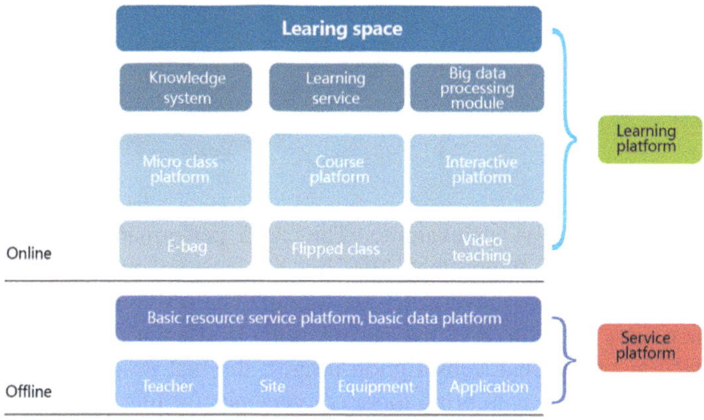

Fig. 8.11 Path to realize smart learning O2O

Offline learning helps learners to validate learning effects. Offline resources shall enable communication between teachers and students. In addition, they can validate the knowledge that has been already acquired., Learners could use the knowledge and skills they get from online learning in some offline contexts to validate the learning effects. Currently, the O2O mode in educational information phase is still under exploration. Business mode and technical mode are not mature; in terms of social vocational education, practice can validate the knowledge that has been learned.

(4) **Cognitive system: the attitude of the general public towards learning should be changed**. Smart learning is an autonomous and immersive learning behavior. It is advocated to learn on demand at any time and any place. Bringing convenience, the Internet also results in discrete and unstructured learning behavior. Most urban citizens have only superficial knowledge about lifelong learning, and usually do not understand the role of fragmentation learning and serious reading in the Internet era. Because of this the use of mobile smart terminals just stays in the entertainment and amusement phase.

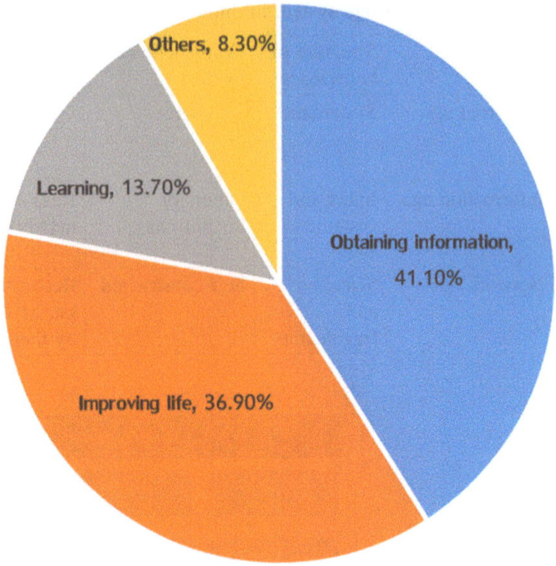

Fig. 8.12 WeChat subscription usage selected by users

Let's take WeChat as an example. The 2015 performance report issued by Tencent in May, 2015 shows that: WeChat has covered more than 90% smart phones, with 549 million active users per month. The institutions for learning and training activities through WeChat spread quickly. In the statistical data about users' selection of WeChat Subscription usages, the first usage is to obtain information. The second one is to bring more convenience for life. Learning knowledge ranks the third, taking up approximately 13.7%[10] (see Fig. 8.12).

Smart learners should well manage themselves. The biggest difference between traditional learning and smart learning is time and space restriction. Traditional education adopts face-to-face teaching mode. This mode has resulted in fixed pattern in learners' minds. Learners are not adopted yet to learning on demand at any time and any place, so their learning is not efficient. Thus, it requires learners to possess strong self-management and autonomous learning ability. Especially, when faced with the vast information and resources provided on the Internet, learners must learn how to avoid deviating from the learning direction.

8.3.3 Mobile Learning Becomes Popular

Mobile network and social media are creating a new learning environment and trigger the second wave of education. On one hand, mobile learning is not a simple migration from PC terminal; on the other hand, mobile devices, such as mobile

[10]Data source: Tencent, 2015 Performance Report, May 2015.

phones and Tablet, realize working, learning and communicating at any time and any place. Meanwhile, in the context of lifelong learning, mobile learning has more profound significance.

(1) **Traditional Internet learning ways present bottlenecks**. "Broadband China" strategy and the fast coverage of wireless network change people's life, learning, and way of communication. According to the Mobile Learning Status Report in China issued by the online education company Skillsoft (2015), 74% learners hold the idea that mobile learning is conductive to obtaining new knowledge and new skills; 54% users express their preference to mobile learning contents. They can view high-quality courses at any time via the hand-held learning platform, select knowledge points to learn based on their own demands, and exchange experiences through social media. Due to such environment and demands, the deficiency of traditional Internet learning platform is increasingly highlighted. It mainly manifests as not rich learning resources, single learning mode, lack of interaction and passive transferring pattern of teaching content, which result in the problem of disconnection between teaching and learning.

(2) **Mobile learning should comply with the evolution of mobile media**. Mobile media as the information output terminal in the mobile Internet era, influences the efficiency of mobile learning in 4 ways. Firstly, the underlying structure of social media is the basic foundation of the content structure of smart learning. Secondly, the organic integration of UGC (user production content) and PGC (professional production content) make UGC serve PGC in a better way. Thirdly, the "we media" of personal learning platform of learners promote the change of learning propagation voice and mode. Fourthly, learning portals are converted into platforms in an accelerated way. The content and service of smart learning are blended.

(3) **Social media promote learning**. It has proved that the favorable combination between social learning and individual learning not only can ensure the diversity, but also can maintain independent thinking. The arrival of social media era strengthens the frequency of communication among people, changes the ways of interpersonal communication and makes the social network mechanisms more effective. It means that social media can improve social learning environment, and help learns absorb more valuable and diversified information.

[Case] Education social network platform of "101.com"

"101.com" created an education social network platform for teachers, students and parents. It helps users obtain resources rapidly via the data pushed actively. Teachers can grasp students' learning progress and effect at any time through this platform. It is convenient for teachers to manage students and conduct targeted teaching; meanwhile, this system is a service platform specially aimed at education, with clear topics and stronger and more convenient interactions among teachers-teachers, teachers-parents and teachers-students.

8.3.4 "Cloud + Client" Fosters the Construction of Smart City

(1) **Understand the "cloud + client" structure in industries**. "Cloud" is the abbreviation for "cloud computing", referring to information infrastructures based on distributed computing architecture, with the capability to cope with big data sets. "Client" refers to terminal devices directly contacted with users, such as the smart phones, PCs, the wearable devices connecting the future Internet of Things (IoT), smart robots, 3D printers and even software and applications. In the critical period of transition from the consumer-oriented Internet to the industry-oriented Internet, all industries, including the education industry, are deeply reformed by the advanced information technologies. Platforms for teaching and learning will be integrated into the "cloud + client" structure.

[Case] Weidong "cloud + client" integration education program

The "cloud" solution of Weidong Cloud Education Group includes the three parts of the cloud computing data center, education resource public service platform and smart interactive class. The "client" solution provides advanced information equipment around high-efficiency interactive class and family-school interconnected writing, as well as hardware support for the innovation of education teaching mode. The "cloud + client" platform will be popularized in 195 countries and regions through UNESCO.

(2) **Cloud computing drives the market popularity of the "education cloud"**. Education cloud includes various resources for the digitization of education, and provides service platforms for education institutions, education staffs and students.[11] Cloud computing integrates the virtualization of data centers, the expansion of bandwidth and the development of intelligent terminals. It contributed to establishing open and shared education service platforms. As the hot spot of current education information, the investment scale of cloud computing has been increasing by years (see Fig. 8.13).

Not only does it consolidate the infrastructure of smart city, but it also lays good foundation for smart education by creating smart learning environment. The popularization and application of education cloud makes educational resources fully utilized, and narrows the urban-rural, region-region and school-school gaps of distribution of educational resources.

(3) **The "dual-core" model promotes the upgrade of smart city**. The "dual-core model of smart city" includes 6 elements, with closer integration with cloud computing, big data and smart terminal. "Smart traffic" relies on the

[11]White Paper on the Development of Cloud Computing Industry (2015 edition)—CCID Think Tank.

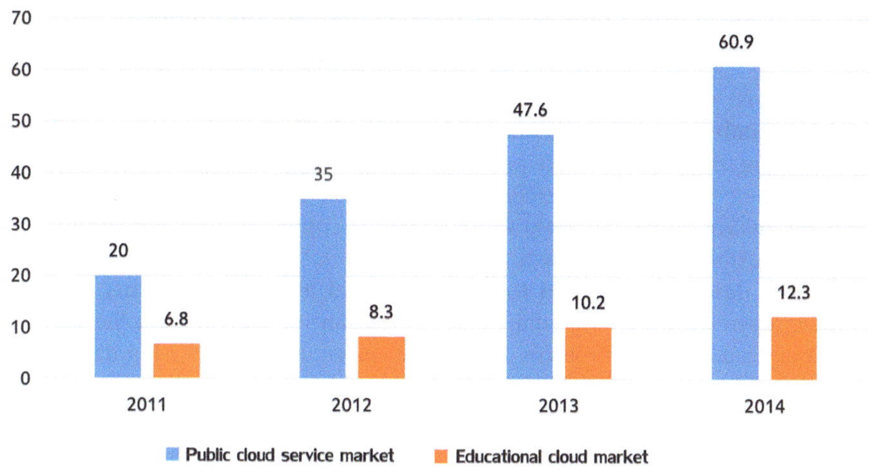

Fig. 8.13 Market size contrast of public cloud and education cloud in 2011–2014

ubiquitous network; "smart learning" needs the mobile terminal application for order of magnitudes; "smart living" relies on the operating system supported by cloud and big data; "smart economy" is the basic path of industrial Internet; "smart governance" depends on the open data; "smart environment" promotes the innovation of civil learning quality.

[Case] "Smart education" conception of Wuxi

Urban construction needs to be led by education. When an "any time, any place, any one and lifelong learning" smart city model becomes gradually utilized, the direction of smart education development will also be defined gradually. *Three-year Action Plan for Education Information of Wuxi (2013–2015)* puts forth efforts on the 6 projects of "public environment support, e-campus construction, digital resources development, lifelong education service, application demonstration for Internet of Things (IoT) and benefaction service improvement". On the basis of family-school communication, educational consultation and emergency communication, it is dedicated to the "smart communication" for family education and education benefaction, and the "smart learning" that provides convenient, flexible and personalized information learning environment for learners.

8.3.5 Use of Big Data to Promote Learning Innovation

Big data is influencing each level of urban system of education. Viktor Mayer-Schönberger said: "big data will make customized education possible in the future." Big data can help to promote learning assessment, learning behavior decision, innovative learning practices, and enhance the level of city services and learning quality. It can also help to reconstruct learning resources system by providing a flexible learning content.

(1) **Big data is aggregated through city-level learning platforms**. Through building open platforms providing lifelong learning opportunities to the citizens, wide-range educational resources can be aggregated, and real-time big data sets of online learning can be established, analyzed and put into application. Thus, it supports the whole life circle of social learning products, from design, development, testing, implementation, evaluation and optimization. Big data promotes evaluation of learning performance, learning behavior decisions, and innovative learning practices Big data also improves urban public service quality. re-constructs learning resources system, and realizes the flexible creation of learning contents. The evolution of smart learning complies with the rule that firstly relying on external drive and then natural evolution. The data-driven feature is the main characteristic in the third phase (see Fig. 8.14).

Urban learning platform is in the exploration phase. According to the investigation on the "citizens' lifelong learning platform in the capital", we found that the user-friendly, interactivity and openness of the platform remained to be

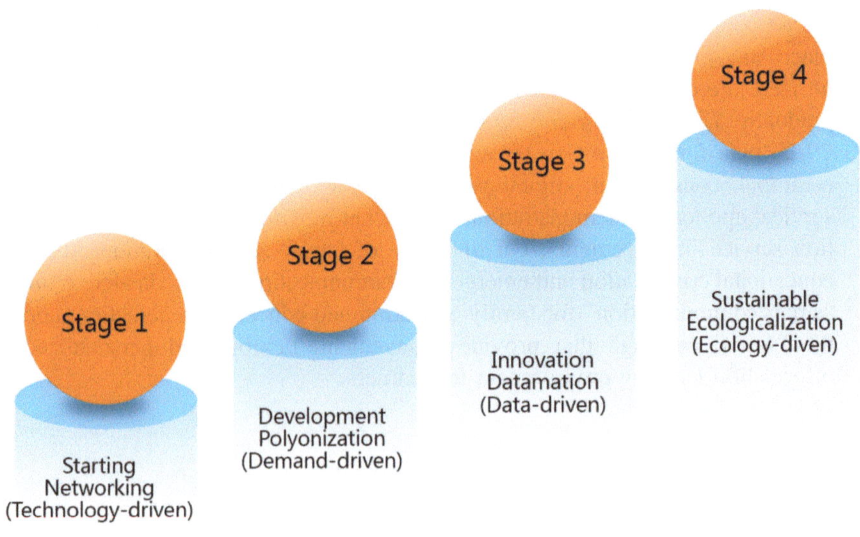

Fig. 8.14 Data-driven personality creation stage

Fig. 8.15 Homepage of Beijing Citizens' lifelong learning platform

improved. In terms of forms, it lacks of links to social media; in terms of contents, it has not realized resource interconnection with education cloud (see Fig. 8.15).

(2) **Big data enables personalized teaching**. In recent decades, personalized teaching is one of the most active fields for education reform. In the new context of teaching and learning, the increasingly diversified technology adapted in education could improve teaching and learning according to learners' difference. Big data effectively avoids the homogenization of learning contents at user interface. Big data enables targeted teaching based on different learning foundations and characteristics of learners.

[Case] Application of learning analysis technology—Khan Academy

There are three factors for the success of Khan Academy—micro-class-type teaching courseware, overturning teaching process and application of learning analysis technology. The platform of Khan Academy provides powerful learning analysis functions. Its teaching activities are divided into the three parts of scholastic achievement, important statistics and community. It prompts "the next recommended learning activity" when entering the learning page at each time. The system can monitor learners' behavior at any time. If teachers find that the teaching video of certain link or knowledge is browsed and clicked by students, it means that such content may be a learning difficulty for students. Thus, teaching shall be adjusted in accordance with it.

Fig. 8.16 Reconstructing educational resource system via big data

(3) **Reconstruct learning resources system by using big data**. The reconstruction of resources can be done in three ways (see Fig. 8.16). Firstly, vitalizing and optimizing resources stock, creating and improving increment of resources, to establish the open resource environment and platform that is available, shareable, exchangeable, applicable, traceable, and optimizable. Secondly, using big data technology, to develop and exploit resource relation value, to increase the cooperative degree and value points among resources, and to realize the precise alignment and high-efficient sharing of resources among environment, participants and platforms. Thirdly, creating tailor-made learning environment and personalized space, to realize oriented target-type pushing. Create and enrich the connotation and practice of smart learning via in-depth cooperative self-organization and social open learning mode, and support the personality creation of learning in the most efficient way.

(4) **Realize the flexible manufacturing of product contents**. Firstly, conduct in-time, precise, dynamic and forward-looking grasp of personalized character and demand of users through using big data. Secondly, according the character and dynamic demand, tailor the product content to realize the cell production, free integration and precision pushing of product contents. Thirdly, make fast response in accordance with the change of internal and external environment, make self-update and adaption, and cater the demand of personality creation to the largest degree. Thus, the interactive and interconnected closed loop was formed among learning demander, product content, and learning provider (see Fig. 8.17).

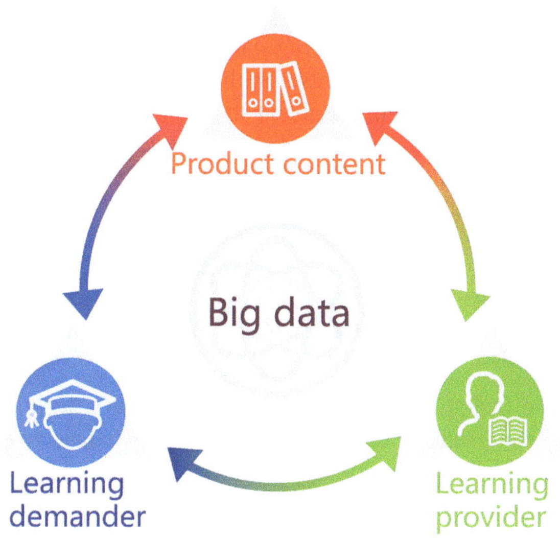

Fig. 8.17 Improving flexibility of product content through big data

8.4 Trends in the Development of Smart Learning

The basic output of smart learning includes learning to learn, learning to do and learning to self-realization. The essence of smart city is to improve social mobilization, social network and social participation, and make members of the society to utilize smart learning as a tool to better comply with social behavior norms.

8.4.1 Integrating the Concepts of Smart Learning and "Internet+"

"Internet+" is a revolutionary concept. The person who firstly changes ideas and seizes the first chance will win opportunities and the power of discourse. The deep integration between "Internet+" and learning is manifested in the improvement of learning infrastructure and the embedding of technology, which result in efficient learning. The change of the image of learning due to thinking and the spirit of the Internet, contributes to the update and upgrade of learning behavior.

Smart learning needs a long-term exploration, and all-round in-depth implementation. It could be implemented in three aspects. With the promotion of education information and lifelong learning society, multiple ideas, subjects and fields will jointly participate in the construction of smart learning. New rules are established through class-level learning platforms, link all people and all industries and form the new ecology of smart learning. Finally, it promotes the fulfillment of smart learning by changing the citizens' mode of learning, production and life.

Smart learning is incorporated into the planning and implementation scope of smart city. Smart learning can provide a large quantity of highly-competent and inter-disciplinary talents for the construction of smart city. It creates the talent cultivation system based on the new demands and new business types for urban construction. It can actively serve the technical innovation for dominant enterprises in cities, innovate disciplinary platform, conduct technical research and development, and strive to make technological breakthrough.

8.4.2 Reconstructing the Structure of Learning

The fusion and development of information technology vanishes the barriers preventing sharing of information and knowledge, as well as the learning boundary. It results in three obvious changes: the first one is the change of instruction methods; the second one is the reconstruction of knowledge system; the third one is the influence on the formation of a new generation of learning way.

Smart terminal application is the partner of the future learning. The Internet of Things and the smart terminal are enhancement organs in the physical world. With the development of the theory, technology and practical application of Internet of Things, human-computer interaction will further upgrade learning experience and learning way. Artificial intelligence, as the powerful support for the future learning, will transit from the self-evolution to the communication with human. It evolves from pure application tools to abilities, and improves learners' wisdom level through the exchange and communication with human.

Mobile learning becomes the paradigm. The link attribute and multi-screen-crossing of the mobile Internet and mobile technology permit people to learn at any time or any place. Mobile, PC and even the TV equipment in living room are combined organically. Learning contents are interactive in multiple platforms and contexts; schools, libraries, classrooms, meeting rooms museums and the public facilities in cities realize knowledge integration through smart perception. Thus, the formal learning and informal learning are interlinked, personal learning and collective learning are harmonious, and class learning and online learning are conducted in a blended mode.

Teaching modes are diversified. In the next 5–10 years, it is predicated that only 15% learners in the world will adopt the traditional learning way, while 85% learners will become the non-traditional learners.[12] Educational institutes provide diversified learning modes that correspond to types of learners; education technical enterprises design the learning situation with multiple learning methods, processes and experiences that satisfy blended learning, online learning and interactive learning, etc. When technology is promoting the high-efficient implementation of

[12]Source: comprehensive predication in accordance with the development of network assistance teaching and online education.

the current learning modes, it will influence community learning and social learning situations profoundly.

8.4.3 Restructuring the Learning Environments

The ubiquitous, ecological and open features of smart learning environment are offsetting the limited, single and closed features of traditional learning environment. The world has been in a turning point from the "model of going to school" to "the mode of surfing the Internet". Learning environment is upgrading from "school-domination" to "network-domination".

Internet makes learning an ability field. The business mode of Internet will deduce towards two directions from pure cash-realization of flow: cloud and big data will be in ascending level; O2O will be in sinking level. In terms of the level of smart learning, Cloud + Client and big data will make industries concentrated towards the direction of platformization. O2O will break through the interaction of educational resources, and lengthen the long tail of application. Smart learning will become people's life instinct in the era of knowledge economy.

Learning contents will become the core of industrial chain. The education in the Internet pattern will present three modes: ground education, Internet platform provider and content provider, etc. The ground education will not disappear, but will be subject to great impact of Internet education. Platform-based enterprises will be the largest network education system in the future; the content providers will be the core of industries. High-quality learning contents are the basic guarantee for the normalization of smart learning.

Vocational education extends to the high-level intellectual groups in cities. With industry upgrading and the aggravation of job market, the highly educated vocational training, and in-service staff vocational training, get more and more attention. Meanwhile, occupational groups have the inborn advantages of higher payment ability and active learning desire. Under the action of policy support and demands, the development of vocational education field will receive much concern.

8.4.4 Reforming the Online Education

Year 2015 was a crucial year for the educational reform in an all-round way in China. The online education filed steps into the phase of a rapid growth. In the next 3–5 years, traditional education and online education will realize the pattern of online taking up 40% and offline taking up 60%.[13] With the sharp advance of the education information, the gap between traditional learning and online learning is

[13]Source: Yu Minhong from the education institution of New Oriental.

decreasing gradually. Learning will develop towards the intelligent, platform-based and smart direction, with more obvious supporting role to smart city.

The cross-border cooperation of online education industry speeds up. The cross-border personnel of online education were faced with a peak season in 2014. On this basis, there will be a large quantity of listed companies and Internet enterprises advancing towards online education in different forms. In the "Internet+" rapid phase, with the release of policies and the constant innovation of Internet enterprises, this segmented field will usher in the period of development opportunity.

Online education becomes the fastest growing field of educational informatization in China. The rapid expansion of online education scale drives users' learning concept and consumption habit, and attracts the attentions of practitioners and VC companies. Through effectively integrating educational resources and Internet technology, the online education service and products, which are highly interactive and encourage personalized learning, are launched frequently. The stickiness of users will be further enhances.

Mobile learning exceeds PC learning. Mobile learning is the main form of online education. In the mobile Internet era, the systematic learning and fragmentation learning are combined organically. Compared to the time of systematic learning, every one possesses more fragmentary time. Smart phones and application software can help people realize the immersive learning. Mobile learning shall pay attention to the depth and breadth of learning. The fragmentation learning shall obtain its effect. The only fragmentation of reading in the fragmentation process shall be changed. By using mobile APPs, people can make good use of fragmentary time, and learn on their own initiative under the rewards mechanism the APPs provide.

8.4.5 Upgrading Smart Learning

With the infusion of technology with education, some characters of technology enhanced learning emerged: online education is combined with offline education; mobile learning is combined with fixed learning; collective learning is combined with individual learning; independent learning is combined with team learning; and knowledge learning is combined with ability cultivation. All these are becoming the new engines of smart learning. The following three characters are also emerging for smart learning.

Game-based mechanism. There have been different views on the game-based learning. On the one hand: games are organically combined with smart learning, because they enhance the student-teacher and student-student interaction and enjoyment of learning in the forms of joint fighting and race. It derives new learning modes and learning experiences. However, on the other hand, game addiction reduces the stickiness of learning.

Video innovation. Technical progress constantly optimizes the learning experiences. The objective of using video in online learning is to provide learners with the experience matching with traditional classroom. Through constant video innovation, the use of video in learning breaks through the inherent thinking paradigm and extends innovative thinking ability.

Bountiful learning experience in classrooms. The interaction mainly manifests the "intelligence" of courseware. The learning courses integrate teaching, learning and exercises. They guide users to actively participate in the learning process, and trigger users to positively think and improve their learning effect; learning contents can be adapted the narrow-band environment; the terse, forceful, lively and interesting "micro class" is popularized quickly.

8.5 Suggestions on How to Promote the Development of Smart Learning

The formation of smart learning is a long-term and continuous process. There are differences in the understanding of the concepts "smart learning in school" and "smart learning in society". We could use "dual-core framework of smart city" and the "framework of smart learning in smart city" proposed in Chaps. 1 and 3 to solve problems emerged in the implementation of smart learning in cities. The following are some of the suggestions for implementing smart learning.

8.5.1 Considering Smart Learning as the New Growth Point of Regional Economic Development

It is necessary to comply with the rules and requirements for high-tech application integration, smart management and modern operation of smart city, and to upgrade the development concept and industrial pattern of smart learning. Firstly, the government should explore the new models to promote the development of smart learning through market mechanism. Secondly, promote enterprises associated with smart learning gradually evolve to an integrated and comprehensive industrial ecology. Thirdly, explore the integration methods of family learning, community learning, social learning, and school learning.

8.5.2 Launching a Wide Range of Pilot Projects

It is required to learn from foreign experiences and practices, to establish smart learning pilot projects in key cities, to promote the smart learning + innovation entrepreneurship. The city should form the learning atmosphere that is harmonious with dual-startup environment, and break through the "linear" learning upgrade pattern in the education. The government of a city could apply new concepts, new methods and new practices of smart learning, obtain low-cost or zero-cost element resources of innovation entrepreneurship, and launch them to markets and users and guide the vitality of cities.

8.5.3 Promoting the Formation of City-Level Learning System

It is required to embed smart learning into the top plans of smart city, and promote it to form city-level learning space system. We should take Internet+ as the bond, advocate the normalization of smart learning, jointly create urban innovation system, motivate innovation vigor and cultivate emerging business type We should also apply modern information technology such the cloud computing, mobile Internet, Internet of Things and big data, accelerate the construction of information infrastructure, meet citizens' learning at any time and any place, and improve resource utilization rate of urban public products and service. In terms of urban layout, we need to pay attention to the cultivation of learning atmosphere, learning situations, and improve citizens' attainment and urban image.

8.5.4 Incorporating Smart Learning Evaluation into the Smart City Evaluation System

Promote the popularity of high-quality education resource to improving citizens' living experience and innovative capability. The current official and third-party smart city evaluation system shall include the assessment of smart learning environment and state. The infrastructure, resource sharing and data application of smart learning should be unified with the big system, platform and ecology of smart city. Meanwhile, we should enlarge policy support to rural, poor, western and minority areas, enlarge the coverage of high-quality resources, and narrow the gap of education between urban and rural areas.

8.5.5 Improving the Governance Mechanism for the Smart Learning

The evolution of rules for smart learning environment indicate that the market, government and society models are faced with different governance dilemmas. Effective governance path lies in the collaboration and interaction among government, market and society. This requires to establish and improve the lifelong learning management system according to urban ecology and objectives, through the guidance of government. The second requirement is to establish the market-driven policy system for smart learning industry, and to realize the benign optimization of smart learning environment relying on industry. The third requirement is to explore the governance mechanism of learning environment with the extensive participation of the public in society, and to mobilize broad social forces to participate in the construction of lifelong learning society.

References

Baidu Library, & Education Management Information Center of Ministry of Education. (2014). *White paper on internet learning in China in 2014 and Research Report on Online Education in 2014*. Retrieved from http://wenku.baidu.com/link?url=cRrXO2rMVXPjpWkSWiDEmjjbvgN6PhGD_7qsIFGtmlSRKYvzNpV7IY5FpOWky-0vPLPmE4_EKeA8BSCDMFaAD2hO65pphc-nkhei9dIMIve

Capital Laboratory. (2015). *Overview of venture investment, merger and acquisition of global and Chinese online education in 2014*. Retrieved from http://www.coinsay.com/article/coinsay_4904.html

China Internet Network Information Center. (2015). *The 35th statistics report on internet development in China*. Retrieved from http://www.cnnic.net.cn/hlwfzyj/hlwxzbg/hlwtjbg/201502/P020150203548852631921

Fudan University. (2014). *Network society mentality report of China*. Retrieved from http://news.sciencenet.cn/htmlnews/2014/10/305476.shtm

Huang, R. (2014). Triple realms of smart education: From environment, model to system. *Modern Distance Education Research, 6,* 3–11.

Huang, R. H., Yang, J. f., & Hu, Y. B. (2012). From digital learning environment to smart learning environment reform and trend of learning environment. *Open Education Research, 1,* 75–84.

HuJiang Online Education. (2005). *The internet learning user behavior report*. Retrieved from http://mt.sohu.com/20150528/n414011755.shtml.2015-5-28

Ministry of Education. (2015). *Statistical bulletin on the development of national educational case in 2014*. Retrieved from http://www.moe.edu.cn/srcsite/A03/s180/moe_633/201508/t20150811_199589.html

Simons, G. (2009). *Knowledge and learning in the network Era—Stepping into the intercommunication*. Shanghai: East China Normal University Press.

Skillsoft. (2015). *Mobile learning situation in China: How mobility influences cloud learning*. Retrieved from http://www.hao5191.cn/news/jiaoyukaoshi/20150909/1229.html

The Central Committee of the Communist Party of China and the State Council. (1996). *The outline for reform and development of education in China (2010–2020)*. Retrieved from http://www.china.com.cn/policy/txt/2010-03/01/content_19492625_3.htm

Chapter 9
Multicultural Perspective on Smart Learning in Smart Cities

Abstract Although we have been able to develop a list of indicators that can measure the "smartness of a city", the vision for each city must be different. It has to be created in a dialog with the local community, and take into consideration cultural and socio-economic context, including cultural traditions, local definition of well being, and governance style. The future of smart cities depends on whether we can educate a new generation of governmental, corporate, and community leaders who will know how to utilize information technology in implementation of said vision. It is important though to consider the fact that just equipping the schools with advanced technology is not enough. As the parameters for smart learning environment will be different for Beijing then for a small city in rural China, we need to build smart schools, with the model of such a school varying for different locations, is necessary. In this chapter we review some opinions on the role of multicultural perspectives on smart cities and smart learning expressed by scientists, government officials teachers, and students. This brief analysis may allow the reader not only to better understand the specificity of Chinese perspective on these concepts, but also the need of teachers training focused on cultural diversity, sustainable development and the role of ICT in governance of the school and smart city.

Keywords Smart cities · Smart schools · China · Cultural diversity · Teachers training · Education · ICT · UNSD goals

9.1 Introduction

As it was mentioned in previous chapters, for the last decade, two concepts became key terms in science, governance, politics, industry and education: sustainability and smartness. In many countries corporations are required to write annual sustainability reports, and sponsoring organizations look for sustainability components in all projects undertaken by local governments and NGO's. The World Commission on Environment and Development defines sustainable development as:

D. Liu et al., *Smart Learning in Smart Cities*, Lecture Notes in Educational Technology, DOI 10.1007/978-981-10-4343-7_9

> …development that meets the needs of the present without compromising the ability of future generations to meet their own needs.

This definition contains two key concepts (see: UN Documents: Our Common Future; Chap. 2: Towards Sustainable Development)

- Concept of 'needs', in particular the essential needs of the world's poor, to which overriding priority should be given; and
- Idea of limitations imposed by the state of technology and social organization on the environment's ability to meet present and future needs.

In 2015 the United Nations published 17 Goals for Sustainable Development as a new pathway for the world community after the Millennium Development Goals era. China declared readiness to participate in this ambitious plan (see the document published in 2015 by the Ministry of Foreign Affairs of the People's Republic of China).

One of the objectives for Goal 9 (Build resilient infrastructure, promote inclusive and sustainable industrialization and foster innovation) says: "*Significantly increase access to information and communications technology and strive to provide universal and affordable access to the Internet in least developed countries by 2020.*" This statement reconfirms that the world already has entered the era of information and communication technology.

In this context another concept became equally popular, namely "smartness." After IBM introduced the concept of the smart city, governments in numerous countries started to promote the idea of building smart cities initially understood as cities incorporating information technology in the governance of the city. Due to increasing energy demands and the ever-growing urban populations, ICT enabled energy efficiency initiatives. Smart Grids, Internet of Things, and other tools have been widely recognized as significant contributors to improving the quality of life in cities. Various aspects of these discussions on the concept of a smart city are discussed in other chapters of this book.

Last year Government of India introduced the 100 Smart Cities for India project. Since then numerous meetings took place during which scientists, members of the governments, and representatives of the corporate business attempted to clarify the vision of smart city, deciding which cites should become smart, and to make plans for implementation of this ambitious initiative (see Smart City Summit New Delhi, 2015, Mumbai 2016). Partnership between governments of USA and India regarding collaboration on 3 smart cities reinforced the efforts.

9.2 Three Perspectives on the Concept of a Smart City

According to Robinson (2013) the first step in creating a smart city is to define what a smart or smarter city means to all stakeholders. Shortly after the birth of the concept of a smart city, most people believed in possibility of creating a general definition that would fit to each and every city around the world. The majority of initial typologies used in classifying smart cities were created from the perspective of the service providers. This perspective is based on two assumptions. The first assumption is that the planners of the city (scientists, corporate business, government officials, politicians, etc.) can create a vision of the city that will be accepted by its inhabitants. The second assumption represents developing and validating a government-centric typology for a smart city that reflects governmental bureaucratic functions, and already existing departments in governmental organizations.

According to Lee and Lee (2014) however, the functions of the smart city should be cross-referenced, so a particular service could be executed across different governmental departments. They are suggesting a citizen-centric typology of a smart city, where the vision of a smart city is developed as a result of the analysis of services needed by the citizens. According to this perspective the vision for a particular smart city must result from a dialog between all stakeholders: citizens, government officials, community organizations, and corporations that will implement this vision.

9.3 Cultural Context and a Vision of Smart City

At one of the international conferences, a community organization reported how they wanted to create a computer center in a remote village in one of the developing countries. They put all the computers on mules and traveled for few days to reach the village. When they arrived they left mules and the computers at the edge of the village, and went to meet with the elders. They told the elders "we have brought you something that will totally change the quality of life in your village. Please go to the edge of the village, and you will find everything there." The elders went to see what was brought, and they took all the mules. That was what they really needed.

This particular situation illustrates why understanding of the cultural context of a city is important for creating of a relevant vision of smartness for this city. The plans for a smart city must always begin with full understanding of the needs of the community, as only in this context one can design a project that will meet needs specific for a given community.

Things, however, may get complicated for a few reasons. First, objectively there may exist a need in a community, yet its members are not really fully aware of it. For example, there may be a need to improve hygiene, as lack of it creates a lot of health issues, but the citizens are not aware of it, as for centuries they have been developing patterns of behavior not meeting our modern health standards.

There is also another important aspect of this situation, and it is illustrated by the following case. A community organization wanted to build a computer center in a village located high in the mountains. They went there with all technology equipment and disappeared for few months. When they came back everybody was asking why they had stayed in this village so long. They answered: "It took us

3 days to teach people how to use computers, but it took us five months to convince the villagers that they need them!"

This example illustrates that even if we bring to the table some commonly accepted standards for quality of life, it may take a lot of educational efforts to convey to members of a particular community that they should follow those standards. Planners of a smart city should first carefully analyze needs of the city and its citizens, and in the case they find needs that are still not recognized by the city, they should work with the community until the message is accepted. This process may become complicated when a community is culturally diversified. With the increasing migration of people, we may have at the same city people with different systems of values, diversified life styles, and understanding of the role of community.

9.4 Cross-Cultural Communication

According to Martin and Beltran (2006) "*Our society is multicultural due to the fact that diverse manifestations of cultural identities live together within it, in addition to it being so due to immigration. Within the society, school education should teach that cultures do not challenge each other but they mutually complement and enrich each other*" (p. 11).

For example, if you visit any city in California, USA, you will find there people from India, Vietnam, Korea, China, and from many African countries. They create their own enclaves, which can be seen as cities within a city. A good example could be "China Towns" in San Francisco or Los Angeles. One can still find people living there who do not speak English, and organize their whole life within the Chinese community speaking only Mandarin. With the new wave of immigrants coming from the Middle East and Africa to Europe, most of the cities in European countries are facing a similar problem.

You may see a big cultural diversity also in many African countries. For example, according to National Geographic in Nigeria only there are more than 250 ethnic groups, including Hausa and Fulani (29%), Yoruba (21%), Ibo (18%), Ijaw (10%), Kanuri (4%), Ibibio (3.5%) and Tiv (2.5%.) The same is in many South American countries, such as the population of Peru, which is composed of Amerindian (45%), Mestizo (37%), White (15%), black, Japanese, Chinese, and other (3%).

What about China? According to Simon (2015) China may appear to outsiders as a country with one language and culture. Such notion is obviously wrong, as China is also a very diverse nation. The Chinese government officially recognizes 56 ethnicities, and many of them have their own language. Besides the dominating Hun ethnicity there are numerous minorities that differ in their customs, names, food preferences, most common sports, etc. Some of the Chinese minorities have cultural traditions rooted in Russia, Korea or Mongolia. Since Deng Xiaoping launched his reforms in 1978, 278 million Chinese workers have moved from villages to cities, greatly accelerating China's growth. Although this trend seems to recently end, as a consequence of this migration China is facing similar problem as cities in USA and Europe. There is a new journal **Cultural Diversity of China** discussing various aspects of this issue, edited by Peter van der Veer and published on the Internet by De Gruyter.

The author of this chapter recalls that at one of the training workshops for facilitators of parent education, organizers presented a video clip illustrating a family meeting of two white American parents with their two teenage children. The issue under discussion was how to avoid constant conflicts between the teens about using clothes that belong to the other sibling, entering the room without notice, etc. The suggested solution was that children should always ask whether they could take something that belongs to the other sibling, and in the case of violating this rule they should pay a $5 fee.

At the workshop there were many participants of Asian origin. One of them, a woman from Korea, stepped out and said: "How dare you to impose on us your bizarre individualistic values; what does it mean, 'this is my shirt, or this is my room'. Everything at home is ours. We are a community, and the only thing we can discuss here is how should we use community property in an organized way." Asian participants warned that if this particular clip and similar ones were not removed for the program, they would simply boycott it.

Community leaders and government officials may face a situation similar to the one described above, when there are conflicting value systems between groups representing different ethnicities. Without understanding differences between cultural backgrounds of community members, community leaders will not be able to bring these people to one table to work together on common issues. According to Robinson (2013) one of the seven steps of developing a smart city is enabling communities to make developing "smartness" a self-sustaining process. However enabling communities that are culturally diversified may become a real challenge for governmental planners, as the process requires understanding the cultural background of all stakeholders, as well as the importance of communication among them.

Elliott et al. (2016) presented a list of 10 myths that prevent collaboration across the cultures. One of them says "*cultural competence is something we each pick up, with time, by working with persons who are different from ourselves*." This is not true, as cultural competence is a skill that requires substantial effort to learn. As the authors indicate, "*working with someone from a different ethnic tradition does not necessarily lead to uncovering differences in expectations, communication styles, and values*." Another myth is that "*an agency should choose a representative from a minority community to represent community's interest to the agency*." This is also not true. According to the authors each community already has a leadership structure, so the agency should rather establish working relationships with elderly in this community, than relay on opinion of a particular member.

9.5 Role of Education in Development of Smart Cities: Concept of Smart Schools

As we may see in discussions on smart cities, participants often refer to the issue of education. To make newly created smart cities sustainable, we have to educate a new generation of governmental officials and community leaders who will understand the vision of a smart city. We need to educate a new generation of businesses managers who will incorporate into their philosophy social responsibility, and the value of sustainable development. Finally, we need to educate a new generation of community members, who as citizens will collaborate with each other and with the government on all issues related to city life. Most importantly though, we need to educate a new generation of global citizens who will understand the issue of cultural diversity, as well as importance of communication between people with different cultural backgrounds.

Introduction of the concept of a smart school was almost a natural consequence of the discussions about the role of information technology in the development of smart cities. The concept of smart schools has an interesting history. In 1984, David Perkins and his colleagues from Harvard University presented an experimental educational program utilizing ICT, known since then as a Project Zero (currently David Wilson is the director of the project and principal investigator). The project is based on two principles:

1. *Learning is a consequence of thinking, and good thinking is learnable by all students.*
2. *Learning should include deep understanding, which involves the flexible, active use of knowledge.*

Initially, it was assumed that it would be enough to create a smart school by equipping schools with computers, and giving to each child a cell phone or tablet. Technology was supposed to give children access to information, and allow communication between students and teachers.

Educationalists quickly realized, however, that although providing students with information is a necessary factor in changing these students' behavior, it is definitely not a sufficient one. All psychologists agree that although obtaining information is a necessary factor for behavioral change, there is much more that needs to follow. First, recipients must believe that the information is true, what relates to the credibility of resources. Then they have to be motivated to apply this information in practice, which usually depends on whether information is compatible with their system of values as this determines importance of information. It also depends on whether they believe that applying of this knowledge will be successful. Cognitive psychologists say "*We don't see things as they are; we see them as we are!*" what very well explains the fact that our reality and value assumptions often determine how we evaluate information. Of course, recipients must also know how to apply this information, and this knowledge must be transformed into skills by practice.

All of the above listed rules apply not only to students, but first and foremost to teachers, who must understand the principles of smart schools and believe that they accurately describe the goals of education. They have to be motivated to implement them, and finally, know how to do it. Here is where knowledge not only about the school subject, but also ICT and developing of relevant practical skills plays a significant role.

In 2010 UNICEF organized a conference in Paris on the Future of Universities, with participation of ministers of education and university presidents from many countries. One of the shared conclusions of the conference was that providing information would no longer be a goal of education. In the era of information technology we need to teach students how to find, evaluate, and apply information.

Motivation is involved at each of these steps. Students have to be motivated to find information (the fact that something is published on the Internet does not mean that everyone knows about it), they must be motivated to evaluate it, and finally, they must have motivation to apply it in everyday life. Simply providing information may not at all affect the motivational system. What is needed is the right learning environment that can appeal to emotions and motivational systems, such as in-service and collaborative learning, interactions with the teachers, relevant feedback, etc.

As we agreed, the main purpose of smart schools is to educate a new generation of leaders and citizens for a sustainable smart city. This notion resulted in the idea that smart schools should not only create the right learning environment and utilize information technology, but also provide clusters of knowledge and skills that traditionally have not been a part of school curriculum, but are important from the perspective of sustainability. As a result of the numerous discussions on smart learning, it was suggested suggest that smart schools should add to their curriculum activities leading to knowledge and skills in the following areas:

- Sustainability and environment;
- Non-violent conflict resolution;
- Ethical governance;
- Global partnerships;
- Vision and development of a smart city;
- Systems approach to complex problems;
- Social work.

These ideas have already found their resonance is a few recently created projects. In the year 2015 the Global Network for Sustainable Development (GNSD), a project developed by the Global Institute of Sustainability at Arizona State University (USA) and the corporation Sustainability Transition Consulting (SusTranCon) presented a proposal for creating a model of a smart school for India. Its implementation, however, met with some challenges. The major one was related to the fact that to make any changes in the curriculum, a school must receive approval from a relevant governmental unit. This process can become very complicated and time consuming. Similar challenge exists in governmental schools in

USA; therefore for the project on our model of smart schools for the USA, we choose the organization of charter schools (Imagine Schools) as they are more independent as regard to curriculum and class activities. In discussions with school principals in India, the alternative solutions to adding new subjects to the curriculum were suggested, namely incorporating the above listed content into already existing curricula for various subjects, or to organizing extra-curricular activities.

An example of such extracurricular activities could be the project Peace Clubs for Sustainability, PCS, initiated by of Mahatma Gandhi Shodh Sansthan (MGSS), a Community Based Organization (CBO) from India. It will be officially launched in New Delhi during the inter-faith and multicultural celebration between the 25th to the 30th of January, 2017. The Peace Club for Sustainability project is a significantly modified version of Sustainability Clubs created by GNSD, as well as by UNESCO Clubs established by UNESCO in schools around Asia and the Pacific. The mission of MGSS is to establish the nonviolent and inclusive society by educating a new generation of community leaders and citizens prepared for building peace in their communities, countries, and the world. Activities of these clubs will introduce students to knowledge about sustainability. Governance, partnerships, vision of smart cities, as well as to the system approach to complex problems.

9.6 Important Lessons from the Malaysia Smart School Project

It is interesting that the most compelling and complex experiment on implementation of the smart school concept has been introduced not in one of the so-called developed countries, but in Malaysia. The Smart School vision came out of a brainstorming session held at the Ministry of Education in 1996. Officials from the MDC, the Ministry of Education and industry representatives produced a Conceptual Blueprint of Smart Schools. They then appointed a consortium of seven Malaysian companies and three multinational companies, in a project management role (please see: Smart School Project Team 1997).

It was the first partnership of its kind for a national education project. In support of this initiative, the Government invested in the development of Malaysia's ICT infrastructure, to enable new technology to be used in the selected schools. Dr. Mahathir Mohamad, one the coordinator of the project, said at lunching of the program:

> We are examining our education system to create a curriculum where people learn how to learn so they can continue their education throughout the rest of their lives. The measure of success in 2020 will be the number and quality of our people who can add value to information.

The Malaysia project defined smart school as "*a learning institution that has been systematically reinvented in terms of teaching-learning practices and school management in order to prepare children for the Information age.*" The assumption was that students should develop in smart schools 21st century skills such as:

- Technology and Media Literacy
- Learning and Innovation Skills
  - Creativity and Innovation
  - Critical Thinking and Problem Solving
  - Communication and Collaboration
- Life and Career Skills.

The main characteristic of a smart school according to Malaysia project should be:

- Student-centered teaching and learning;
- Students that exhibit higher order thinking skills;
- Teachers and Administrators who are skillful in using ICT in daily tasks;
- Teachers who are innovative and creative in using ICT as an enabler and accelerator for better teaching and learning;
- Schools that have smart partnerships with various agencies.

The project has been introduced into 4 phases (see: Bahagian Teknologi Pendidikan, Kementerian Pelajaran Malaysia 1996).

- Pilot phase (1999–2002) conducted in 88 schools selected nationwide;
- Post-Pilot Phase (2002–2005) during which massive computerization to all 10,000 school was introduced;
- Making All Schools Smart (2005–2010) schools were leveraging all ICT initiatives;
- Consolidation and Stabilization (2010–2020) all innovative practices using ICT are implemented.

Malaysia's Smart School project involves a wide range of inter-related initiatives. These include schemes to improve Malaysia's ICT infrastructure, training in change management for teachers and school managers, a national school management system to link schools and the communities they serve, integration of software, and a help desk facility (see: ICT in Education, UNESCO Bangkok).

We have to agree that the Malaysia Project is very ambitious and attempts to involve all potential stakeholders that could contribute to its success. As the final evaluation of the project is expected after the year 2020, we do not have too many evaluations of the impact it has had on schools. One of the early reports says that the result of the incorporation of ICT into schools is at a rate not far behind the rates for more developed nations (Kader 2008).

Another evaluation written by Majeed and Yusoff (2015), represents a more reserved opinion. Authors attempted to evaluate what ICT tools are available at the school and to what extent do the teachers use ICT in teaching and learning, in addition to the factors that contribute to the teachers' use of ICT in teaching. The school that was a subject of their evaluation has been a part of Smart School Project for 15 years. The results of their study confirm, in our opinion, the concerns that simply equipping schools with technology tools is not enough to reach aimed goals.

The authors conclude "*that in spite of being a smart school for the past 15 years, the participants' ICT adoption was still at a low level. This could be due to the fact that extended training was not provided regularly. In addition, the participants felt that the school management did not put ICT adoption as a priority and that the time allocated to use and explore ICT tools as well as prepare the resources was not enough. Even though the participants' attitudes were positive and ICT infrastructure was in place, it was still not sufficient to cater to the needs of the teachers and students…Therefore, the level of ICT adoption in the smart school in this study was still at a basic level, which is the Utilization phase perhaps due to the barriers that the participants highlighted. Directly, this clearly reflects that the school has not yet achieved the smart school status as projected*" (pp. 255–56).

One of the recommendations made by Majeed and Yusoff is that the number and type of training programs provided to the teachers should be increased, as only this could improve and encourage the teachers' adoption to the new strategy of teaching and learning.

Our assumption is that some cultural factors should be involved. In an interesting study Teo et al. (2008), made a cross-cultural examination of intention to use technology between Singaporean and Malaysian teachers. What they found that although there was no difference between the groups in the area of behavioral intention, the groups differed in the area of perceived usefulness of ICT, as well as in attitudes towards computers and the ease of using them. These results indicate that before implementing ICT in educational process we need to carefully evaluate the starting point to know teachers' and students' exact attitudes and expectations. If necessary, and before attempts to implement technology, we should then organize the process of training intending to change the attitudes.

This process, however, can be very long, as it was in the case of the organization trying to create a computer center in the remote village. It seems that in attempts to motivate people for changing of their behavioral patterns it is important to consider the fact that most people have a natural resistance to change. As the change requires acquiring new patterns of behavior, or inhibiting prior ineffective habits, during the transitional phase from the old patterns to the new ones, the latter usually require more effort so people are less affective. This phenomenon is well described and explained by Bridge's Transition Model (Bridges 2009) suggesting how people should be guided through the process of change. Definitely students and teachers from different cultural and economic backgrounds may require a shorter or longer process of transition from the traditional way of teaching and learning, to the one incorporating ICT.

9.7 Smart Learning Environment

As it was discussed in previous chapters, smart learning environment can be regarded as the technology-supported learning environment that make adaptations and provides appropriate support (e.g., guidance, feedback, hints or tools) based on individual learners' needs (Huang et al., 2012). Typically we associate the concept of smart learning environment with a smart school.

The most commonly quoted model of a smart school called Project Zero was founded at Harvard University in 1967 by David Perkins and has been running until today. The team of co-workers (see the website Project Zero) presented 7 characteristics of a smart school that in a sense indicate that smart schools should provide smart learning environment. Here are some of those characteristics:

Generative knowledge. *Schools must examine carefully what disciplinary and interdisciplinary content will most benefit students. Identifying and structuring content that has the greatest potential for students' development is an important starting point for the Smart Schools model.*

Learnable intelligence. *Contrary to a psychological tradition that tends to view intelligence as a fixed quantity, much of the research of Project Zero and others' indicates that students can and do learn ways of thinking that can boost their performance. The integration of the teaching of higher order thinking into subject matter instruction and the creation of a school culture that champions and scaffolds such thinking can have a significant effect on students' own views of their abilities and on their learning.*

Focus on understanding. *While there are many legitimate goals for students, often a focus on deep understanding gets lost in the day-to-day life of the school. In the Smart Schools model, we place an emphasis on student work that builds and demonstrates deep understanding in contrast to rote or narrowly defined outcomes.*

Teaching for mastery and transfer. *A simple but powerful maxim of education is that students learn much of what they have a reasonable opportunity and motivation to learn. Teaching techniques that explicitly model, scaffold, motivate, and help students to bridge what they learn to new contexts (i.e., transfer) greatly enhance the likelihood that students will learn well and actively use what they learn.*

Learning-centered assessment. *Assessment at its best, functions as a reflective and evaluative tool for learning. It involves students as well as teachers and creates a dynamic in which students take on the ultimate responsibility for the quality of their work and their learning.*

Embracing complexity. *Insightful thinking and deep understanding require students to be able to deal with and even thrive on complex situations and problems. The Smart Schools model involves learning situations that help students build skills and tolerance for complexity and begin to develop a sense of excitement in the face of intriguing and difficult problems. It also supports teachers in managing the complexities of new viewpoints and practices.*

The school as a learning organization. *Just as schools are places of growth for children, they should be places of growth for faculty and administrators—places where the pursuit of intellectual interests and professional collaborations are supported and encouraged. In addition, the successful learning organization institutes structures that enable all members of the school community to collaborate in the processes of direction-setting and self-monitoring, creating a dynamic system that changes as the needs and the vision of the community changes.*

Interestingly enough, although ICT is not directly mentioned in any of those principles, in a different part of the project the authors clarify their view on this issue:

> Digital technologies offer exciting opportunities for learning, civic engagement, and intercultural exchange, among other vital areas of life. Virtual worlds and mobile devices offer new ways of learning and transfer of learning beyond the classroom. Online communities can connect young people from diverse backgrounds to exchange perspectives, while informal learning and civic engagement opportunities abound on social media sites, blogs, and other online spaces.

The Zero Project also mentions that it is really vital that educators will leverage the positive affordances of digital media and support youth to use them in reflective positive ways, as the Internet can also introduce distractions, misinformation and negative interactions.

R. Huang, the Dean of the Smart Learning Institute recently created at the Beijing Normal University, during the press conference organized in 2015 presented a white paper "Smart Learning Environment in 2015 China" (Huang 2015) in which he introduced ten key concepts for developing smart learning environment in China. Below are some of them:

(1) *A smart learning environment is the foundation of learning in the age of information. A smart learning environment enables people to learn at any time and at any place, in their own styles and at their own paces. This environment supports learning more easily, attentively and effectively.*
(2) *The establishment and development of a smart city operates on the "double engines" of "livability experience" and "innovative energy." A smart learning environment contributes significantly to the "livability experience."*
(3) *With the advancement of the smart city concept and increased educational opportunity in our society, learning at home, in the community and at work will become more important and will become critical components of lifelong learning, along with learning at school.*
(4) *"Smart learning" should be the core for the building of the smart city. It supports lifelong learning and can serve as a critical feature for an urban system's "self-evolution." The development of "smart learning" can enhance a city's livability experience and vitalize a city's creativity, thus uncovering the "smartness" of a city.*

After reading about these key concepts we must agree that in the same way as impact of culture has to be considered in the process of development of a smart city, it must be taken into consideration in organization of educational efforts. A lot of research reports on the influence of culture on learning and motivation have been published at the beginning of this century. We strongly recommend to all educators the book "Culture, Motivation and Learning: A Multicultural Perspective" edited by Salili and Hoosain (2007).

According to the authors, the aim of their book is: "*...to present research findings and views of scholars and researchers in the field of motivation and learning, from a multicultural and international perspective. Educators and scholars from different parts of the world have examined recent learning and motivation theories in different cultural contexts in order to explore the dynamics of sociocultural processes affecting student motivation. Others have focused on teaching and learning strategies that are known to be effective with culturally diverse students.*"

9.8 Impact of Cultural Context on Students' Motivation for Learning

Need for Achievement. According to Maehr and Braskamp (1986) and Salili (1994), students with different cultural backgrounds attach different values and meanings to achievement, and therefore they approach achievement tasks in a different way. In one of Salili's cross-cultural studies comparing Chinese and British high school students, it proved that although the dimensions of achievement were similar across the two cultures, the meanings students attached to achievement and the ways they achieved their goals were significantly different. Academic achievement was significantly more important for the Chinese than British students, while career was significantly more important for the British. In another more recent study by Salili et al. (2001), they compared Chinese immigrants to Canadian students and found that goals of pleasing parents, friends, and teachers were rated significantly more important by Chinese students then by Canadian students (pp. 7–8).

Cooperative versus Competitive Learning. More and more educators agree that the traditional teacher-oriented curriculum no longer meets the needs of multicultural schools (see for example Nieto 2004). When we look at the characteristics of smart schools as described previously, they put a strong accent on collaborative and cooperative learning as the tool developing independent thinking and problem solution skills. According to Garton (2007), learning is a socio-cultural event that can be achieved only in a collaborative context. Competition and collaboration are often seen as mutually exclusive, the first one being typical for individualistic cultures while the second characterizes collectivistic cultures. Interestingly enough, some authors suggest that competition can be used to motivate students in collaborative and group work (see Salili and Hoosain 2007, p. 13; Fulop et al. 2007).

9.9 Literacy: Traditional and New Definitions

In our discussion on the importance of technology based education we cannot neglect the statistics provided by UNESCO:

- *Today, nearly 17% of the world's adult population is still not literate; two thirds of them women, making gender equality even harder to achieve.*
- *The scale of illiteracy among youth also represents an enormous challenge; an estimated 122 million youth globally are illiterate, of which young women represent 60.7%.*
- *The 67.4 million children who are out of school are likely to encounter great difficulties in the future, as deficient or non-existent basic education is the root cause of illiteracy. With some 775 million adults lacking minimum literacy skills, literacy for all thus remains elusive.*

These numbers create a big challenge for the visionaries of smart cities and smart schools. They also indicate that providing relevant infrastructure must back up the process of building motivation for learning. For readers who had an opportunity to visit some of the governmental schools in India as well as some private schools in this country, it is obvious how dramatic the differences are between these school in the aspect of infrastructure, staffing, and supply of instructional materials. There are, however, new definitions of literacy introduced nowadays. For example, The Colorado Department of Education defines technology literacy as the ability to appropriately select and responsibly use technology. Students who have attained technological literacy are able to use information technology to:

- Solve problems
- Communicate
- Locate, use and synthesize information
- Develop skills necessary to function in the 21st century.

Again, it looks that smart schools should meet these goals. Children who are attending these often neglected schools are at the least at school and receive basic education, so we can still discuss what would the obligations be of smart cities towards those schools, and how to create in these schools a smart learning environment.

9.10 Conclusion

Although we did not cover in this chapter the entire area of cultural diversity and multicultural education, we hope to have presented some of arguments for the need to consider the impact of cultural context in the planning of a smart city. The common phenomenon of migration, both on the national and global scales, creates a challenge for the government of smart cities, for community leaders, and for communities that must coexist within the city. They need to understand cultural diversity and have basic skills allowing multicultural communication. As we have mentioned above, for the vision of smart cities to succeed we need to educate a new generation of governmental leaders, community leaders and citizens who will be co-creators of a smart city.

Here is where the role of smart schools is supposed to create smart learning environments. These schools should prepare students for their roles in a smart city. To assure positive effects of these processes teachers need to first acquire knowledge and skills enabling them to effectively use information technology in the educational process. Another important element of this preparation should again be knowledge about cultural diversity and multicultural communication. To be effective in this task teachers at smart schools must understand the cultural background of their students, and organize school activities in a way reflecting cultural differences between them. This seems to be a necessary condition for developing motivation for learning and building partnerships across cultural borders.

Finally, all the reflections presented above lead us to the conclusion that any hope that creating smart classrooms equipped with the sophisticated technology will bring more effective learning may become just a wishful thinking if we neglect one extremely important step. Namely, incorporation of technology must be followed by the intense training of both teachers and students intended to create new attitudes and change in their traditional patterns of behavior.

References

Bahagian Teknologi Pendidikan, Kementerian Pelajaran Malaysia. (1996). *The Malaysian smart schools: Developing 21st century skills.* http://www.jaet.jp/katudou/seminar_ict/Malaysia.pdf

Bridges, W. (2009). *Managing transitions: Making most of change.* Lifelong Books.

China's Position Paper on the Post 2015 Development Agenda. (2015). *Ministry of Foreign Affairs of the People's Republic of China.* http://www.fmprc.gov.cn/mfa_eng/wjdt_665385/2649_665393/t1263455.shtml

Cultural Diversity; Universal Declaration of UNESCO. In *31st Meeting of General Conference, Paris*, November 1, 2001

Elliott, C., Adams, R. J., & Sockalingam, S. (2016). *Ten myths that prevent collaboration across cultures.* http://www.awesomelibrary.org/multiculturaltoolkit-myths.html

Fulop, M., Ross, A., Kuscer, M. P., & Pucko, C. R. (2007). Competition and cooperation in schools: An English, Hungarian, and Slovenian comparison. In F. Salili & R. Hoosain (Eds.), *Culture, motivation, and learning* (pp. 235–284). Charlotte, NC: Information Age Publishing.

Garton, A. F. (2007). Understanding the effect of culture on the pursuit of multiple goals, learning through collaboration: Is there a multicultural perspective? In F. Salili & R. Hoosain (Eds.), *Culture, motivation, and learning: A multicultural* (pp. 195–216). Charlotte, NC: Information Age Publishing.

Global Institute of Sustainability (GIOS) at Arizona State University. (n.d.). https://sustainability.asu.edu/

Global Network for Sustainable Development (GNSD). (n.d.). http://gnsd.org

Huang, R. (2015). *Smart learning environment in 2015 China.* http://english.bnu.edu.cn/academicevents/79456.htm

Huang, R., Yang, J., & Hu, Y. (2012). From digitalized environment to smart environment: the reform and trends of learning environment. *Open Education Research, 18*(1), 75–84.
Imagine Schools. (n.d.). http://www.imagineschools.org/
Kader, B. K. B. A. (2008). *Malaysian Smart Schools project*. ICT in Education, UNESCO Bangkok. http://www.unescobkk.org/fr/education/ict/online-resources/databases/ict-in-education-database/item/article/malaysian-smart-schools-project/
Lee, J., & Lee, H. (2014, June). Developing and validating a citizen-centric typology for smart city services. *Government Information Quarterly, 31,* S93–S105. http://www.sciencedirect.com/science/article/pii/S0740624X14000598. doi:10.1016/j.giq.2014.01.010
Maehr, M. L., & Braskamp, L. A. (1986). *The motivation factor: A theory of personal investment*. Lexington, MA: Lexington Books.
Majeed, Z. S., & Yusoff, Z. S. (2015). *Journal of Studies in Education, 5*(3). ISSN 2162-6952. See also https://www.researchgate.net/publication/281618344_Are_We_'Smarter'_Now_Case_Study_of_Smart_School_Implementation_in_a_Developing_Nation
Martin, B. S., & Beltran, J. M. (2006). *Multiculturalism and immigration*. http://www.lasalle.org/wp-content/uploads/pdf/mel/cahier_mel/26cahier_mel_en.pdf
Nieto, S. (2004). *Affirming diversity: The sociopolitical context of multicultural education* (4th ed.). Boston: Pearson Education Inc.
NigeriaFacts. (n.d.). http://travel.nationalgeographic.com/travel/countries/nigeria-facts/
Our Common Future. UN Documents: Report of the World Commission on Environment and Development. (n.d.). http://www.un-documents.net/ocf-02.htm
Project Zero. (n.d.). http://www.pz.harvard.edu/projects/smart-schools
Robinson, R. (2013). *7 steps to smart city*. https://theurbantechnologist.com/seven-steps-to-a-smarter-city/
Salili, F. (1994). Age, sex, and cultural differences in the meaning and dimensions of achievement. *Personality and Social Psychology Bulletin, 20*(6), 635–648. doi:10.1177/0146167294206002
Salili, F., Chiu, C., & Lai, S. (2001). The influence of culture and context on students' motivational orientation and performance. In *Student motivation* (pp. 221–247). doi:10.1007/978-1-4615-1273-8_11
Salili, F., & Hoosain, R. (2007). *Culture, motivation and learning: A multicultural perspective*. Information Age Publishing, Inc.
Simon, G. E. (2015). *How diverse is China in terms of culture, language, etc.?* https://www.quora.com/How-diverse-is-China-in-terms-of-culture-language-etc
Smart Cities Summit Mumbai. (2016). http://www.asappconferences.com/smartcities/index.aspx
Smart School Project Team. (1997). The Malaysian Smart School Project; A conceptual blueprint. Government of Malaysia (pp. 1–177). http://www.mscmalaysia.my/sites/default/files/pdf/publications_references/Smart_School_Blueprint.pdf
Sustainability Transition Consulting, LLC (SusTranCon). (n.d.). http://www.sustrancon.org
Teo, T., Luan, W. S., & Sing, C. C. (2008). A cross-culural examination of the intention to use technology between Singaporean and Malaysian pre-service teacher: An application of the Technology Acceptance Model (TAM). *Educational Technology & Siociety, 11*(4), 265–280.
Tolan, C. (2014). *Cities of the future? Indian PM pushes for 100 smart cities*. CNN http://www.cnn.com/2014/07/18/world/asia/india-modi-smart-cities/
UN Sustainable Development Goals. https://sustainabledevelopment.un.org/?menu=1300
van der Veer, P. (2015). *Cultural diversity in China*. http://www.mmg.mpg.de/publications/cultural-diversity-in-china/

MIX
Papier aus verantwortungsvollen Quellen
Paper from responsible sources
FSC® C105338

If you have any concerns about our products,
you can contact us on
ProductSafety@springernature.com

In case Publisher is established outside the EU,
the EU authorized representative is:
Springer Nature Customer Service Center GmbH
Europaplatz 3, 69115 Heidelberg, Germany

Printed by Libri Plureos GmbH
in Hamburg, Germany